Lecture Notes of the Institute for Computer Sciences, Social Informatics and Telecommunications Engineering 577

The LNICST series publishes ICST's conferences, symposia and workshops.
LNICST reports state-of-the-art results in areas related to the scope of the Institute.
The type of material published includes

- Proceedings (published in time for the respective event)
- Other edited monographs (such as project reports or invited volumes)

LNICST topics span the following areas:

- General Computer Science
- E-Economy
- E-Medicine
- Knowledge Management
- Multimedia
- Operations, Management and Policy
- Social Informatics
- Systems

Jiageng Chen · Zhe Xia
Editors

Blockchain Technology and Emerging Applications

Third EAI International Conference, BlockTEA 2023
Wuhan, China, December 2–3, 2023
Proceedings

 Springer

Editors
Jiageng Chen
Central China Normal University
Wuhan, China

Zhe Xia 🅾
Wuhan University of Technology
Wuhan, China

ISSN 1867-8211 ISSN 1867-822X (electronic)
Lecture Notes of the Institute for Computer Sciences, Social Informatics
and Telecommunications Engineering
ISBN 978-3-031-60036-4 ISBN 978-3-031-60037-1 (eBook)
https://doi.org/10.1007/978-3-031-60037-1

Preface

We are happy to introduce the proceedings of the 2023 European Alliance for Innovation (EAl) International Conference on Blockchain Technology and Emerging Applications (BlockTEA), held in Wuhan 2nd–3rd December 2023.

Motivated by the success of cryptocurrency, blockchain technology has been emerging as a technology with potential applications in various domains, including finance, computer science, electronic engineering, agriculture, healthcare and more. The blockchain-based emerging applications are able to aid current systems and networks by leveraging the benefits provided by blockchain technology, such as a decentralized, immutable and cryptographically secured ledger. The aim of BlockTEA is to bring together researchers and practitioners from different disciplines to discuss recent advancements in blockchain and its emerging applications.

The technical program for BlockTEA 2023 featured 8 full papers presented orally during the main conference sessions, giving an acceptance rate of approximately 25.8%. We received a total of 31 submissions, each of which underwent rigorous double-blind review by three high-quality reviewers. Aside from the high-quality technical paper presentations, the technical program also featured four keynote speakers: Mingwu Zhang from Hubei University of Technology, Yining Liu from Guilin University of Electronic Technology, Zhe Xia from Wuhan University of Technology and Chi Cheng from China University of Geosciences. The titles of the four talks are "Privacy Preservation on Machine Learning", "Privacy issues in data publication", "Quorum Controlled Homomorphic Re-encryption for Privacy Preserving Computations in the Cloud" and "On key reuse attacks against lattice-based KEMs".

Coordination with the steering chairs, Imrich Chlamtac and Weizhi Meng, was important for the success of the conference. We sincerely appreciate their constant support and guidance. It was also a great pleasure to work with such an excellent organizing committee team for their hard work in organizing and supporting the conference. In particular, the Technical Program Committee, led by our General Co-Chair Mingwu Zhang (Hubei University of Technology), and all the other chairs contributed to the conference organization and made a high-quality technical program. We are also grateful to all the authors who submitted their papers to the conference.

We strongly believe that the BlockTEA conference provides a unique platform for all researchers, developers and practitioners to discuss all state-of-the-art and recent advancements that are relevant to Blockchain Technology and Emerging Applications. We also expect that future BlockTea conferences will be as successful and stimulating as indicated by the contributions presented in this volume.

December 2023

Jiageng Chen
Zhe Xia

Organization

Steering Committee

Imrich Chlamtac   University of Trento, Italy
Weizhi Meng   Technical University of Denmark, Denmark

Organizing Committee

General Chair

Weizhi Meng   Technical University of Denmark, Denmark

General Co-chair

Mingwu Zhang   Hubei University of Technology, China

TPC Chairs and Co-chairs

Jiageng Chen   Central China Normal University, China
Zhe Xia   Wuhan University of Technology, China

Sponsorship and Exhibit Chair

Javier Parra-Arnau   Universitat Politècnica de Catalunya, Spain

Local Chairs

Pei Li   Central China Normal University, China
Shixiong Yao   Central China Normal University, China

Workshops Chair

Jun Shao   Zhejiang Gongshang University, China

Publicity and Social Media Chairs

Kambourakis Georgios	University of the Aegean, Greece
Jiangshan Yu	Monash University, Australia

Publications Chair

Wenjuan Li	The Hong Kong Polytechnic University, China

Web Chair

Wei-Yang Chiu	Technical University of Denmark, Denmark

Technical Program Committee

Jiageng Chen	Central China Normal University, China
Zhe Xia	Wuhan University of Technology, China
Owen Arden	University of California, Santa Cruz, USA
Man Ho Au	University of Hong Kong, China
Xinxin Fan	IoTex, USA
Xiapu Luo	Hong Kong Polytechnic University, China
Angelo De Caro	IBM Zurich, Switzerland
Jian Liu	Zhejiang University, China
Nikos Lenonardos	University of Athens, Greece
Yajin Zhou	Zhejiang University, China
Alberto Sonnino	Facebook Novi, UK
Ghassan Karame	NEC Laboratories, Germany
Pei Li	Central China Normal University, China
Shixiong Yao	Central China Normal University, China
Ralph Holz	University of Twente, The Netherlands
Alessandro Sorniotti	IBM Research, Switzerland
Bhaskar Krishnamachari	University of Southern California, USA
Ali Dorri	Queensland University of Technology, Australia
Christof Ferreira Torres	University of Luxembourg, Luxembourg
Wenjuan Li	Hong Kong Polytechnic University, China
Andrea Bracciali	University of Stirling, UK
Foteini Baldimsti	George Mason University, USA
Adam Everspaugh	Station70, USA
Ittay Eyal	Technion – Israel Institute of Technology, Israel

Contents

Applied Cryptography and System Security

Identity-Based Key Verifiable Inner Product Functional Encryption Scheme

Mingwu Zhang[1,2]([✉]), Chao He[1], and Gang Shen[1]

[1] School of Computer Science, Hubei University of Technology, Wuhan, China
csmwzhang@gmail.com
[2] School of Computer Science and Information Security, Guilin University of
Electronic Technology, Guilin, China

Abstract. Functional encryption is a novel form of public key encryption that has captured significant attention since its inception, with researchers proposing a series of theoretical constructions. Functional encryption can be investigated for specific real-world applications such as the evaluation and output from the ciphertexts using the different decryption keys. In this paper, we investigate one of the more popular recent developments in functional encryption, i.e., inner product functional encryption. We address potential issues that inner product functional encryption might encounter in certain scenarios, including the inability to specify the identity of the ciphertext recipient, privacy leakage related to the master secret key vector, and the susceptibility of the decryption key to malicious tampering. In specific contexts, there might be a requirement for ciphertext recipients to be carefully designated. Malicious adversaries holding the decryption key can exploit it to gain insight into the master key or even alter the decryption key information. Consequently, key verification becomes necessary. To address this, we propose an identity-based key verifiable inner product functional encryption scheme (IBVE-IPE), which can effectively resolve the aforementioned issues and is validated for security and practicality through security proofs and performance analyses.

Keywords: Inner Product Functional Encryption · Identity-Based · Key Verification

1 Introduction

Functional encryption (FE) is a multi-function cryptographic primitive that was first formalized by Boneh [9]. In conventional public-key encryption, decryption functions as an all-or-nothing process, wherein the recipient can either fully obtain the information using the decryption key or obtain nothing at all. Conversely, functional encryption, a novel approach in public-key encryption, allows for precise control over the extent of information disclosed by a ciphertext to a specific recipient. Since the introduction of functional encryption, many

J. Chen and Z. Xia (Eds.), BlockTEA 2023, LNICST 577, pp. 3–22, 2024.
https://doi.org/10.1007/978-3-031-60037-1_1

researchers have been considering how to implement generalized constructions of FE [8,16]. Early works used more complicated theoretical tools, such as Indistinguishability Obfuscation (IO) and Multilineal Map, which are of doubtful practicality. Hence, constructing a specialized and efficient FE scheme tailored to the requirements of specific applications has become a compelling area of exploration for numerous researchers. In this paper, we delve into one of the more prevalent forms of FE in recent times: inner product functional encryption (IPFE). Our aim is to address the challenges associated with IPFE, particularly its inability to specify the recipient's identity within the ciphertext and the potential issues of master key privacy leakage and key tampering in specific scenarios.

We consider that inner product functional encryption is widely used in various scenarios, and we are aware that different functions are required for schemes in different contexts, which may limit its applicability in specific scenarios. The following issues have been taken into consideration regarding inner product functional encryption schemes in particular scenarios. Therefore, we present a new functional encryption scheme to overcome the following challenges.

1) Unable to specify identity of recipient: In certain practical application scenarios, limitations may arise when the encrypting party cannot specify a recipient for the ciphertext. For instance, consider an outsourced database hosted on a cloud service, containing data ciphertexts uploaded by the data owner. Supposing the data owner intends to solely offer data usage to paying users. If it's impossible to specify the recipient of the ciphertexts, there would be no way to prevent non-paying users from accessing the database. Without the ability to screen the identity of users accessing the database, there is a lack of control over who can access it. This situation would be unfair to the paying user that is subscribed to the service.

2) Master secret key privacy leakage: The master secret key vector leakage problem of applying inner product functional encryption in outsourcing described in work [18]. The key generation center generates the master secret key vector $s = (s_1, s_2, ..., s_n)$. The decryptor can keep querying the key generation center for the decryption key sk_y associated with the vector $y = (y_1, y_2, ..., y_n)$, with sk_y reflecting the inner product of s with y. Due to the linear nature of the inner product, if a malicious decryptor queries decryption keys $sk_{y^{(1)}}, ..., sk_{y^{(n)}}$ for a set of n vectors $y^{(1)}, ..., y^{(n)}$ that are linearly independent, they can derive a system of linear equations

$$\begin{cases} sk_{y^{(1)}} = \sum_{i=1}^{n} s_i y_i^{(1)} \\ \quad \cdots \\ sk_{y^{(n)}} = \sum_{i=1}^{n} s_i y_i^{(n)} \end{cases} \tag{1}$$

Solving this system of linear equations allows them to obtain the value of the master secret key vector s.

3) Decryption key modification attack: Considering that there are relevant nearest neighbor query services in real-world application scenarios, such as the

application of secure k-nearest neighbor algorithms for mining and querying data in cloud environments. Essentially, in article [17], it is indicated that the k-nearest neighbor (KNN) query can be represented as an inner product computation between two vectors. If an adversary intercepts and modifies the query key of the querying user, it eventually impacts the querying user's query results. For example, in certain existing inner product functional encryption schemes, the secret key form is expressed as $SK_y = (\boldsymbol{y}, \langle \boldsymbol{s}, \boldsymbol{y} \rangle)$, where $\boldsymbol{y}$ represents the query vector and $\boldsymbol{s}$ denotes the master key vector. If the adversary intercepts l keys $SK_{y_i} = (\boldsymbol{y}_i, \langle \boldsymbol{y}_i, \boldsymbol{s} \rangle)_{(i \in (1,...,l))}$. It can compute to generate $\boldsymbol{y}^* = k_1 \boldsymbol{y}_1 + ... + k_l \boldsymbol{y}_l$ and $\langle \boldsymbol{s}, \boldsymbol{y}^* \rangle = k_1 \langle \boldsymbol{s}, \boldsymbol{y}_1 \rangle + ... + k_l \langle \boldsymbol{s}, \boldsymbol{y}_l \rangle$, where $k_1, ..., k_l$ are integers chosen by the adversary. If the adversary transmits the altered secret key to cloud server for decryption, the decryption result returns as $\langle \boldsymbol{x}, \boldsymbol{y}^* \rangle$ instead of the original $\langle \boldsymbol{x}, \boldsymbol{y} \rangle$.

1.1 Related Work

In 2015, Abdalla *et al.* [3] introduced a novel encryption concept known as inner product functional encryption (IPFE), within the public-key cryptography. In their paper, they developed particular, effective strategies relaying on well-established assumptions like the DDH and LWE [13] assumptions, and these strategies offer security against Chosen Plaintext Attacks (CPA) within the selective model. In 2016, they also analyzed various security notions for functional encryption for inner product evaluation in [4] and presented a new generic construction that provides security against adaptive adversaries. They demonstrated this construction through three examples based on different assumptions. Subsequently, Agrawal *et al.* [7] presented constructions that provably achieves secure adaptive attacks on the same inner product function. Additionally, they devised a approach based on Paillier's [12] composite residual assumption, which facilitates inner product evaluation for short integer vectors. This enabled inner product evaluation modulo the RSA [14] integer N. Goldwasser *et al.* [11] presented the concept of Multi-Input Functional Encryption (MIFE), in which a secret key SK_f can be associated with an n-ary function f capable of processing multiple ciphertexts as its input. Some of the work in recent years has considered multi-input and multi-client scenarios, Abdalla *et al.* [6] proposed an innovative approach to convert single-input functional encryption designed for inner products into multi-input schemes with equivalent functionality. This method did not involve pairing and could be deployed with any single-input schemes, and proposed the initial function-hiding MIFE scheme for inner products relying on standard assumptions. Chotard *et al.* [10] combined the techniques of Private Stream Aggregation [15] and Functional Encryption to introduced a primitive called Decentralized Multi-Client Functional Encryption (DMCFE). Abdalla *et al.* [2] introduced a labeled DMCFE scheme secured under the plain DDH assumption without pairings. They also provided universal compilers that convert any method from the less robust model, in which the adversary must request access to the encryption oracle for every ciphertext generated by each party, into a scheme without this restriction. In another work, Abdalla *et al.* [5]

explored Multi-Client Functional Encryption schemes that support encryption labels. These labels enabled the decryptor to decrypt ciphertexts with the same label, thus restricting the potential for mix-and-match during decryption. They addressed limitations found in previous works [1,10] by introducing three buildings featuring ciphertexts of linear size. These constructions were built upon the Matrix-DDH (MDDH), DCR [12] and LWE assumptions within the framework of the random oracle model. Additionally, Yang *et al.* [18] constructed an effective outsourced inner product computation scheme utilizing IPFE. They raised the issue of master key vector leakage, through their inner product scheme lacks an authenticated function. Nevertheless, the data privacy of the data owner remains preserved on the untrusted cloud server. Zhang *et al.* [19,20] extended the concept of IPFE to the domain of neural network prediction and machine learning. Their work applied MCFE to scenarios where multiple data providers collaborated to predict labels and implemented privacy preservation of word vector training, respectively.

1.2 Our Contributions

Within this work, we present an identity-based key verifiable inner product functional encryption scheme (IBVE-IPE) that can obtain the leakage resilience of master key and the decryption verification. In particular, the primary contributions can be condensed into the following three facets:

- At first, our IBVE-IPE scheme enables the specification of the identity of the ciphertext recipient. That addresses the issue of being unable to designate the ciphertext recipient, which is a common problem in various scenarios, and it caters to the needs of specific practical application scenarios.
- Secondly, our scheme effectively prevents the leakage of master secret key information. Additionally, the decryption key can be verified, which thwarts attempts at malicious tampering. This feature enhances privacy preservation.
- Thirdly, our scheme is designed to meet the demands that require specified identities and privacy preservation. While there is a performance overhead trade-off, it remains at an acceptable level. The security proofs and performance evaluations provide evidence that our scheme integrates security and practicality.

1.3 Organization

The structure of the remaining sections of this paper is organized as shown below. Section 2 provides an introduction to the essential preliminaries. We outline the scheme's definition and the security model in Sect. 3. We present our construction in Sect. 4. Section 5 provides the security analysis and essential proofs. Section 6 offers the presentation of the performance analysis. The conclusion of this work is in Sect. 7.

2   Preliminaries

Within this phase, we introduce the notations and some pertinent cryptography that we use in this paper.

2.1   Notation

Here we intention present the symbols utilized in this work. Let λ denote the security parameter in a cryptographic system. Let n denote the message length parameter. Let X, Y be the boundary parameters of vectors $\boldsymbol{x}, \boldsymbol{y}$, respectively. Let N be the safety modulus. Let mpk and msk denote the master public key and the master secret key of the system, respectively. Let H represent a full-domain cryptographic hash function. Let $\boldsymbol{x}$ represent a message vector to be encrypted. Let $\boldsymbol{y}$ denote the decryption vector. Let ID denote the specified identity information. Let sk_y denote the decryption key. Let pk_y denote the public key of decryptor. Let $\langle \boldsymbol{x}, \boldsymbol{y} \rangle$ represent the inner product between vectors $\boldsymbol{x}$ and $\boldsymbol{y}$. Let $\mathcal{A}$ denote adversary and $\mathcal{C}$ denote challenger in the security model.

2.2   Cryptographic Primitives

Definition 1 (DCR Assumption). Provided with an integer N that is the product of prime numbers p and q, the Decision Composite Residuosity (DCR) problem, as described in [12], involves distinguishing between two distributions: $\mathcal{D}_0 = \{z | z \leftarrow \mathbb{Z}_{N^2}^*\}$ and $\mathcal{D}_1 = \{z | z = z_0^N \bmod N^2, z_0 \leftarrow \mathbb{Z}_N^*\}$, and the advantages probability $Adv_{\mathcal{A}}^{DCR}(\lambda)$ of distinguishing is negligible.

Definition 2 (Inner Product Functional Encryption Based on DCR Assumption). This definition of IPFE is quote from the construction based on DCR assumption in [7].

- **Setup**$(1^\lambda, 1^n, X, Y)$: Given a security parameter λ, a length parameter n and two bound parameters X, Y. Select secure prime numbers p and q, where $p = 2p' + 1$ and $q = 2q' + 1$, with $p', q' > 2^{l(\lambda)}$ for a polynomial l. Calculate the composite modulus N as $N = pq$, ensuring $N > XY$. Randomly select $g' \leftarrow \mathbb{Z}_{N^2}^*$ and calculate g as $g = g'^{2N} \bmod N^2$. Select an integer vector $\boldsymbol{s} = (s_1, \ldots, s_n) \leftarrow \mathcal{D}_{\mathbb{Z}^n, \sigma}$, where $\sigma > \lambda^{\frac{1}{2}} \cdot N^{\frac{5}{2}}$. Calculate the public key components $(h_i = g^{s_i} \bmod N^2)_{i \in [1,n]}$. Output the public key

$$mpk = (g, (h_i = g^{s_i} \bmod N^2)_{i \in [1,n]}, N, X)$$

  and keep the secret key $msk = (\boldsymbol{s}, Y)$.
- **KeyGen**$(msk, \boldsymbol{y})$: Given a vector $\boldsymbol{y} = (y_1, \ldots, y_n) \in \mathbb{Z}^n$ with $\|\boldsymbol{y}\| \leq Y$ and msk, compute and output the decryption key $sk_y = \sum_{i=1}^n s_i \cdot y_i$ over $\mathbb{Z}$.
- **Encrypt**$(mpk, \boldsymbol{x})$: Given a vector $\boldsymbol{x} = (x_1, \ldots, x_n) \in \mathbb{Z}^n$ with $\|\boldsymbol{x}\| \leq X$ and mpk, then select a random number $r \leftarrow 0, \ldots, \lfloor N/4 \rfloor$ to compute

$$C_0 = g^r \bmod N^2,$$

$$C_i = (1 + x_i N) \cdot h_i^r \ mod \ N^2, \forall i \in \{1, ..., n\}.$$

Return $C_x = (C_0, C_1, \ldots, C_n) \in \mathbb{Z}_{N^2}^{n+1}$.
- **Decrypt**(mpk, sk_y, C_x): Given $sk_y \in \mathbb{Z}$, mpk and C_x, compute

$$C_y = (\prod_{i=1}^{n} C_i^{y_i}) \cdot C_0^{-sk_y} \ mod \ N^2$$

Subsequently, calculate $\frac{C_y - 1 \ mod \ N^2}{N}$ and return the result $\langle x, y \rangle$.

3 IBVE-IPE Definition and Security Model

This section provides an overview of the IBVE-IPE scheme, including its definition and security model.

3.1 Definition of IBVE-IPE

Definition 3 (IBVE-IPE). IBVE-IPE scheme is comprised of five algorithms, i.e., IBVE-IPE = (**Setup, KeyGen, Encrypt, Verify, Decrypt**). The description of each algorithm is provided below:

- **Setup**$(1^\lambda, 1^n, X, Y)$: Input the security parameter λ, the length parameter n, and the boundary parameters X, Y. As a result of the algorithm, it produces the key pair mpk and msk.
- **KeyGen**(msk, ID, y): Input msk, the identity ID of the key holder, and the vector $y \in \mathbb{Z}^n$ with $\|y\| \leq Y$. It produces the decryption key sk_y and the public key pk_y as outputs.
- **Encrypt**(mpk, pk_y, ID, x): Input mpk, pk_y, the ciphertext receiver identity ID, and the vector to be encrypted $x \in \mathbb{Z}^n$ with $\|x\| \leq X$. The algorithm outputs the ciphertext C_x.
- **Verify**(sk_y, mpk, y): Input sk_y, mpk, and the vector y. The algorithm outputs 1 to indicate that the key verification passed, i.e., the decryption key has not modified. Otherwise, if it outputs 0, it means that the verification failed, i.e., the key has been changed.
- **Decrypt**(mpk, C_x, sk_y): Input mpk, ciphertext C_x, and the key sk_y. If the ciphertext C_x and decryption key sk_y correspond to the same ID, it yields the result $\langle x, y \rangle$ as output.

3.2 Security Model of IBVE-IPE

Definition 4 (Adaptive Security of IBVE-IPE). In order to establish the adaptive security of IBVE-IPE, we initially present the framework outlined in the game denoted as $Exp_{\mathcal{A}}^{Adv}$, which involves the interaction of $\mathcal{A}$ with $\mathcal{C}$.

- **Setup:** $\mathcal{C}$ executes **Setup**$(1^\lambda, 1^n, X, Y)$ to produce (mpk, msk) and subsequently transmits mpk to $\mathcal{A}$.

- **Query phase 1:** $\mathcal{A}$ adaptively selects the challenge identity ID^* and vectors $\boldsymbol{y}^{(1)}, ..., \boldsymbol{y}^{(l_1)}$ for $\mathcal{C}$. Subsequently, $\mathcal{C}$ utilizes **KeyGen**$(msk, ID^*\boldsymbol{y}^{(i)})$ to generate secret keys $sk_{\boldsymbol{y}^{(1)}}, ..., sk_{\boldsymbol{y}^{(l_1)}}$ and public keys $pk_{\boldsymbol{y}^{(1)}}, ..., pk_{\boldsymbol{y}^{(l_1)}}$ for $\mathcal{A}$.
- **Challenge:** $\mathcal{A}$ adaptively selects two vectors $\boldsymbol{x}^{(0)}$ and $\boldsymbol{x}^{(1)}$ under the condition that $\langle \boldsymbol{x}^{(0)}, \boldsymbol{y}^{(i)} \rangle = \langle \boldsymbol{x}^{(1)}, \boldsymbol{y}^{(i)} \rangle$ holds for all $i \in \{1, ..., l_1\}$ for $\mathcal{C}$. Subsequently, $\mathcal{C}$ selects a random bit $\mu \in_R \{0, 1\}$, and invokes **Encrypt**$(mpk, pk_{\boldsymbol{y}^{(i)}}, ID^*, \boldsymbol{x}^{(\mu)})$ for one $i \in \{1, ..., l_1\}$ to produce the challenge ciphertext $C_{\boldsymbol{x}^{(\mu)}}$ intended for $\mathcal{A}$.
- **Query phase 2:** After obtaining $C_{\boldsymbol{x}^{(\mu)}}$, $\mathcal{A}$ adaptively selects vectors $\boldsymbol{y}^{(l_1+1)}, ..., \boldsymbol{y}^{(l_1+l_2)}$ under the condition that $\langle \boldsymbol{x}^{(0)}, \boldsymbol{y}^{(i)} \rangle = \langle \boldsymbol{x}^{(1)}, \boldsymbol{y}^{(i)} \rangle$ for every $i \in \{l_1 + 1, ..., l_1 + l_2\}$ for $\mathcal{C}$. $\mathcal{C}$ executes **KeyGen**$(msk, ID^*, \boldsymbol{y}^{(i)})$ to produce the secret keys $sk_{\boldsymbol{y}^{(l_1+1)}}, ..., sk_{\boldsymbol{y}^{(l_1+l_2)}}$ and the public keys $pk_{\boldsymbol{y}^{(l_1+1)}}, ..., pk_{\boldsymbol{y}^{(l_1+l_2)}}$ for $\mathcal{A}$
- **Guess:** $\mathcal{A}$ provides its guess μ'. If and only if $\mu' = \mu$ then show that $\mathcal{A}$ wins this game.

The IBVE-IPE scheme achieves adaptive security when $\mathcal{A}$ attains a success probability of $\frac{1}{2} + \epsilon$ in $Exp_{\mathcal{A}}^{Adv}$, where ϵ is a negligible quantity.

Definition 5 (Master Secret Key Hiding of IBVE-IPE). In order to establish the concept of concealing the master secret key in the IBVE-IPE scheme, we begin by introducing the framework described in the game $Exp_{\mathcal{A}}^{MSKH}$ involving $\mathcal{A}$ and $\mathcal{C}$.

- **Setup:** $\mathcal{C}$ executes the **Setup**$(1^\lambda, 1^n, X, Y)$ to produce (mpk, msk) and subsequently transmits mpk to $\mathcal{A}$.
- **Query:** $\mathcal{A}$ adaptively selects the challenge identity ID^* and vectors $\boldsymbol{y}^{(1)}, ..., \boldsymbol{y}^{(l)}$ for $\mathcal{C}$. Subsequently, $\mathcal{C}$ generate $sk_{\boldsymbol{y}^{(1)}}, ..., sk_{\boldsymbol{y}^{(l)}}$ and $pk_{\boldsymbol{y}^{(1)}}, ..., pk_{\boldsymbol{y}^{(l)}}$ by constantly running **KeyGen**$(msk, ID^*, \boldsymbol{y}^{(i)})$ for $\mathcal{A}$.
- **Guess:** $\mathcal{A}$ provides its guess msk', and $\mathcal{A}$ is declared the winner in this game if and only if $msk' = msk$.

The IBVE-IPE scheme achieves concealing the master secret key if the probability of $\mathcal{A}$ winning in $Exp_{\mathcal{A}}^{MSKH}$ is ϵ, where ϵ is negligible.

Definition 6 (Identity Key Attack Security Model of IBVE-IPE). To define the identity key attack security model of IBVE-IPE, we begin by introducing the framework described in the game $Exp_{\mathcal{A}}^{IKA}$ involving $\mathcal{A}$ and $\mathcal{C}$.

- **Init:** $\mathcal{A}$ selects the challenge identity ID^* and the vector $\boldsymbol{y}^*$ and sends them to $\mathcal{C}$. $\mathcal{A}$ aims to generate a new key $sk_{\boldsymbol{y}'}^{(ID^*)}$, with $\boldsymbol{y}' \neq \boldsymbol{y}^*$
- **Setup:** $\mathcal{C}$ runs **Setup**$(1^\lambda, 1^n, X, Y)$ to generate key pair (mpk, msk) and subsequently transmits mpk to $\mathcal{A}$.
- **Query:** $\mathcal{A}$ is permitted to request l keys $sk_{\boldsymbol{y}^i}^{(ID^i)}(i \in [1, l])$ from $\mathcal{C}$, where the constraint is $ID^i \neq ID^j (i, j \in [1, l])$. Furthermore, the challenge identity ID^* must be bound to the challenge vector $\boldsymbol{y}^*$.

- **Forge**: $\mathcal{A}$ outputs the forged key $sk_{y'}^{ID^*}$, where $y' \neq y^*$.

The advantage of $\mathcal{A}$ winning the $Exp_{\mathcal{A}}^{IKA}$ game we define as $Adv_{IKA} = Pr[\mathcal{A}(mpk, \forall_{(i\in[1,l])} sk_{y^i}^{(ID^i)}(ID^i \neq ID^*), sk_{y^*}^{(ID^*)}) = sk_{y'}^{(ID^*)} : y' \neq y^*]$. If $\mathcal{A}$'s advantage Adv_{IKA} is negligible, we can argue that IBVE-IPE is resistant to identity key modification attacks.

4   Our Construction

Within this section, we present a comprehensive explanation of our scheme. We refer to the construction based on Paillier in [7] to construct our method. In this scheme, we consider key vectors y and message vectors x, with an assumption that their magnitudes are bounded: $\|x\| \leq X$ and $\|y\| \leq Y$. These specific bounds, denoted as X and Y, are deliberately selected to ensure that their product, $X \cdot Y$, remains less than the modulus N of Paillier's cryptosystem. By adhering to these norm constraints, the decryption process allows for the retrieval of $\langle x, y \rangle$ modulo N, which is equivalent to $\langle x, y \rangle$ in integer terms. As such, we accept the premise that both X and Y fall within the range of values less than $(N/n)^{1/2}$, while retaining the original meaning and context.

- **Setup**$(1^\lambda, 1^n, X, Y)$: Given a security parameter λ, a length parameter n and two boundary parameters X, Y. Select secure primes $\tilde{p} = 2p' + 1$ and $\tilde{q} = 2q' + 1$, with p' and q' are suitably large prime numbers exceeding a threshold defined by a polynomial function $2^{l(\lambda)}$. Calculate the composite modulus $N = \tilde{p}\tilde{q}$, ensuring that N surpasses the product $X \cdot Y$. Randomly select $g' \leftarrow \mathbb{Z}_{N^2}^*$ and calculate $g = g'^{2N} \bmod N^2$, which yields a subgroup consisting of $(2N)$th residues within $\mathbb{Z}_{N^2}^*$ with a high likelihood. Pick a hash function $H : \{0,1\}^* \rightarrow \mathbb{Z}_N^*$ with collision resistant. Then, sample a number $t \in \mathbb{Z}_N^*$ that has the inverse under the Euler's totient function of N^2 and an integer vector $s = (s_1, ..., s_n) \leftarrow \mathcal{D}_{\mathbb{Z}^n, \sigma}$, where $\sigma > \lambda^{\frac{1}{2}} \cdot N^{\frac{5}{2}}$, from which to compute $h = g^t \bmod N^2$ and $(h_i = g^{s_i} \bmod N^2)_{i \in (1,...,n)}$. Let

$$mpk = (g, N, H, h, \{h_i\}_{i=1}^n, X)$$

and $msk = (t, s, Y)$.
- **KeyGen**(msk, ID, y): Input msk, ID and a vector $y = (y_1, y_2, ..., y_n) \in \mathbb{Z}^n$ with $\|y\| \leq Y$. Select $\alpha \in_R \mathbb{Z}_N$ and calculate

$$sk_1 = \langle s, y \rangle + t\alpha + H(ID)$$

$$sk_2 = \alpha + t^{-1} H(ID)$$

$$pk_y = h^\alpha \bmod N^2$$

Then sets $sk_y = (sk_1, sk_2)$ as decryption key and make pk_y public.

- **Encrypt**$(mpk, pk_y, ID, \boldsymbol{x})$: Input mpk, pk_y, ID and a vector $\boldsymbol{x} = (x_1, x_2, ..., x_n) \in \mathbb{Z}^n$ with $\|\boldsymbol{x}\| \leq X$, choose a random value $r \leftarrow 0, ..., \lfloor N/4 \rfloor$ and calculate

$$\sigma = H(g^{H(ID)r}), C_1 = g^r \ mod \ N^2$$

$$C_2 = pk_y^r \ mod \ N^2$$

$$C_3 = \{ct_i = h_i^r \cdot (1 + N\sigma x_i) \ mod \ N^2\}_{i \in [1,n]}$$

Return $C_x = (C_1, C_2, C_3, ID)$.
- **Verify**$(mpk, \boldsymbol{y}, sk_y)$: Input mpk, $\boldsymbol{y}$ and key sk_y. If the subsequent equation is true, then produce the output 1:

$$(\prod_{i=1}^{n} h_i^{y_i}) \cdot h^{sk_2} \ mod \ N^2 \overset{?}{=} g^{sk_1} \ mod \ N^2$$

Otherwise, output 0.
- **Decrypt**(mpk, C_x, sk_y): Input mpk, sk_y and ciphertext C_x. If the identity ID in C_x does not match with that in sk_y, then the algorithm outputs $\bot$ and aborts. Conversely, the corresponding decryption computations proceeds as following.

$$\omega_1 = C_1^{H(ID)} \cdot C_2 \cdot (\prod_{i=1}^{n} ct_i^{y_i}) \cdot C_1^{-sk_1} \ mod \ N^2$$

$$\omega_2 = \frac{(\omega_1 - 1)(H(C_1^{H(ID)}))^{-1} \ mod \ N^2}{N} = \langle \boldsymbol{x}, \boldsymbol{y} \rangle$$

Finally return decryption result $\langle \boldsymbol{x}, \boldsymbol{y} \rangle$.

5   Security Analysis

Within this section, we delve into our scheme, including consideration of its correctness and security.

5.1   Correctness of IBVE-IPE

The fundamental principle of correctness revolves around the idea that if the operations are carried out faithfully within the system, the ultimate outcome achieved by the decryptor corresponds to $\langle \boldsymbol{x}, \boldsymbol{y} \rangle$. The presented IBVE-IPE scheme is indeed correct. In a formal manner, we present the ensuing theorem.

Theorem 1. *The presented IBVE-IPE scheme is correct.*

Proof: If all participants perform the procedures of the program honestly, then we can derive the subsequent outcomes:

$$(\prod_{i=1}^{n} h_i^{y_i}) \cdot h^{sk_2} \bmod N^2 = g^{\langle s,y \rangle} \cdot g^{t(\alpha+t^{-1}H(ID))} \bmod N^2$$

$$= g^{\langle s,y \rangle} \cdot g^{t\alpha+H(ID)} \bmod N^2 \qquad (2)$$

$$= g^{\langle s,y \rangle+t\alpha+H(ID)} \bmod N^2$$

$$= g^{sk_1} \bmod N^2$$

$$\omega_1 = C_1^{H(ID)} \cdot C_2 \cdot (\prod_{i=1}^{n} ct_i^{y_i}) \cdot C_1^{-sk_1} \bmod N^2$$

$$= g^{r(H(ID)+t\alpha+\langle s,y \rangle)} \prod_{i=1}^{n} (1+N\sigma x_i)^{y_i} \cdot g^{-r \cdot sk_1} \bmod N^2 \qquad (3)$$

$$= (1+N\sigma \sum_{i=1}^{n} x_i y_i) \bmod N^2$$

$$\omega_2 = \frac{(1+N\sigma \sum_{i=1}^{n} x_i y_i - 1)(H(g^{rH(ID)})^{-1}) \bmod N^2}{N}$$

$$= \frac{(N\sigma \sum_{i=1}^{n} x_i y_i)\sigma^{-1} \bmod N^2}{N} \qquad (4)$$

$$= \frac{(N \sum_{i=1}^{n} x_i y_i) \bmod N^2}{N}$$

$N \sum_{i=1}^{n} x_i y_i$ remains less than N^2 due to our initial assumptions at the commencement of Sect. 4, where we stipulate that x and y satisfy $||x|| \le X$, $||y|| \le Y$ and $X \cdot Y \le N$. Consequently, we can express ω_2 as follow:

$$\omega_2 = \frac{N \sum_{i=1}^{n} x_i y_i}{N} = \sum_{i=1}^{n} x_i y_i$$

Therefore, the proposed IBVE-IPE scheme is correct.

5.2   The Adaptive Security of IBVE-IPE Scheme

The security analysis of the IBVE-IPE scheme demonstrates that, when provided a ciphertext of vector x, the sole piece of content disclosed by the secret key concerning vector y is $\langle x, y \rangle$. More precisely, given the adaptively selected secret keys $sk_{y(1)}, ..., sk_{y(l_1+l_2)}$ and the situation that $\langle x^{(0)}, y^{(i)} \rangle = \langle x^{(1)}, y^{(i)} \rangle$ for all $i \in \{1, ..., l_1 + l_2\}$, it remains impossible for $\mathcal{A}$ to differentiate whether the encrypted vector is $x^{(0)}$ or $x^{(1)}$. The adaptive security framework grants $\mathcal{A}$ permit to the system shared parameters and permits queries for a succession of secret keys before selecting the challenge vectors $x^{(0)}$ and $x^{(1)}$. The implementation of adaptive security is fundamental feature of our IBVE-IPE scheme. Formally, this is corroborated by the subsequent theorem.

Theorem 2. *The IBVE-IPE scheme is adaptive security based on the DCR assumption.*

Proof: The adaptive security of the IBVE-IPE scheme is established through a succession of games, initiating with a scenario in which the adversary receives with a genuine encryption of $x^{(\mu)}$, where μ is a randomly chosen bit from $\{0,1\}$, and culminating in a situation in which the value of μ becomes statistically independent of the $\mathcal{A}$'s observations. For each iteration, we represent the event of the adversary winning as S_i.

Game 0: It represents a truly secure game. The adversary $\mathcal{A}$ is provided with $mpk = (g, N, H, h, \{h_i\}_{i=1}^n, X)$, where $h = g^t \bmod N^2$ and $h_i = g^{s_i} \bmod N^2$. In the challenge phase, $\mathcal{A}$ selects challenge identity ID^* and two vectors, denoted as $x^{(0)}$ and $x^{(1)}$, where $||x^{(\mu)}|| < X$, and acquires an encryption $C_{x^{(\mu)}}$ for a single bit $\mu \in \{0,1\}$. At the conclusion of the game, $\mathcal{A}$ produces μ' based on the knowledge at hand. We use S_0 to represent the event where $\mu' = \mu$. Once again, it is necessary for the equality $\langle x^{(0)}, y \rangle = \langle x^{(1)}, y \rangle$ to hold true for any vector y offered by the adversary to the oracle.

Game 1: This game we adjust the encryption information $C_{x^{(\mu)}} = (C_1, C_2, C_3, ID^*)$ of the vector $x^{(\mu)}$. To be precise, the challenger $\mathcal{C}$ selects $\eta_0 \in_R \mathbb{Z}_{N^2}$ and calculates

$$\eta = \eta_0^N \bmod N^2 \qquad (5)$$

Subsequently, utilizing $msk = (s, t, Y)$, it proceeds to calculate the ciphertext components just as it did in *Game* 0. The remaining parts are determined in the subsequent manner:

$$C_1 = \eta^2 \bmod N^2, C_2 = C_1^\alpha \bmod N^2 \qquad (6)$$

$$C_3 = \{ct_i = C_1^{s_i}(1 + N \cdot \sigma)^{x_i^{(\mu)}} \bmod N^2\}_{i \in [1,n]} \qquad (7)$$

The distribution of $C_{x^{(\mu)}}$ closely that of *Game* 0, given that C_1 is now entirely uniform within the subgroup of $(2N)$th residues. Therefore, we can express $|Pr[S_1] - Pr[S_0]| \leq \frac{1}{2^\lambda}$.

Game 2: In the challenge phase, we once more adjust how $C_{x^{(\mu)}} = (C_1, C_2, C_3, ID^*)$ is generated. This implies that in place of deriving C_1 by initially picking a random Nth residue η from $\mathbb{Z}_{N^2}^*$, the challenger now randomly selects $\eta \leftarrow \mathbb{Z}_{N^2}^*$ and computes C_1 following the method employed in *Game* 1. Then the rest of the ciphertext parts are calculated as in *Game* 1. Under the presumption that the DCR assumption holds, such alterations do not wield substantial influence on $\mathcal{A}$, leading to the conclusion that

$$|Pr[S_2] - Pr[S_1]| \leq Adv_{\mathcal{A}}^{DCR}(\lambda) \qquad (8)$$

Game 3: It closely resembles *Game* 2, with the sole difference being how the message $x^{(\mu)}$ is encrypted compared to *Game* 2. $\mathcal{C}$ selects $a_\eta \in_R \mathbb{Z}_N^*$ and $r_\eta \in_R \{0, ..., \lfloor \frac{N}{4} \rfloor\}$, then calculates $C_1 = g^{r_\eta} \cdot (1 + N \cdot a_\eta) \bmod N^2$. And the remaining ciphertext parts are calculated following the same procedure as in *Game* 2. The statistical difference between the distribution of C_1 in *Game* 3 and *Game* 2 is less then $\frac{1}{2^\lambda}$. Thus, we can conclude that

$$|Pr[S_3] - Pr[S_2]| \leq \frac{1}{2^\lambda} \tag{9}$$

In *Game* 3, for every $j \in \{1, ..., n\}$

$$ct_j = h_j^{r_\eta} (1 + N \cdot \delta)^{(x_j^{(\mu)} + a_\eta \cdot s_j) \bmod N} \bmod N^2 \tag{10}$$

Later, we will demonstrate that the Eq. (10) effectively conceals the randomly chosen bit $\mu \in_R \{0, 1\}$. Therefore, to substantiate the assertion made earlier, we analyze the vector

$$\boldsymbol{x} = (x_1, ..., x_n) = \frac{1}{k}(\boldsymbol{x}^{(1)} - \boldsymbol{x}^{(0)}) \tag{11}$$

in which $k = gcd(x_1^{(1)} - x_1^{(0)}, ..., x_n^{(1)} - x_n^{(0)})$. We are aware that due to the inherent constraints of the inner product functional encryption, all legitimate key queries pertaining to the vector $\boldsymbol{y} \in \mathbb{Z}^n$ must adhere to the conditions within the lattice $\{\boldsymbol{y} \in \mathbb{Z}^n | \langle \boldsymbol{x}, \boldsymbol{y} \rangle = 0\}$, where the rows of the matrix $Y_{top} \in \mathbb{Z}^{(n-1) \times n}$. Furthermore, we know $||\boldsymbol{x}||^2 < N$ and we can reasonably presume that $gcd(||\boldsymbol{x}||^2, N) = 1$. Otherwise, the simplification process could potentially determine a non-trivial factor of N. Let us define the matrix

$$\boldsymbol{Y} = \begin{pmatrix} Y_{top} \\ \boldsymbol{x} \end{pmatrix} \in \mathbb{Z}^{n \times n} \tag{12}$$

Due to the condition $||\boldsymbol{x}||^2 < N$, the matrix $\boldsymbol{Y}$ is invertible in $\mathbb{Z}_N$, except when its determinant exposes a non-trivial factor of N.

For all $j \in [1, n]$, the Eq. (10) content discloses the subsequent vector in theory:

$$\begin{aligned} \boldsymbol{w}^{(\mu)} &= ((x_1^{(\mu)} + a_\eta \cdot s_1) \bmod N, ..., (x_n^{(\mu)} + a_\eta \cdot s_n) \bmod N) \\ &= ((\boldsymbol{x}^{(\mu)} + a_\eta \cdot \boldsymbol{s}) \bmod N) \end{aligned} \tag{13}$$

Hence, it is imperative to demonstrate that, from $\mathcal{A}$'s perspective, $\boldsymbol{w}^{(\mu)}$ exhibits independence with respect to $\mu \in \{0, 1\}$. As $\boldsymbol{Y} \in \mathbb{Z}^{n \times n}$ is non-dependent on $\mu \in \{0, 1\}$ and remains invertible over $\mathbb{Z}_N$, this allows us to establish the statistical independence of $\boldsymbol{Y} \cdot (\boldsymbol{w}^{(\mu)})^T \in \mathbb{Z}_N^n$ with respect to $\mu \in \{0, 1\}$. Considering that $Y_{top} \cdot (\boldsymbol{x}^{(1)} - \boldsymbol{x}^{(0)}) = 0$, it's evident that $Y_{top} \cdot (\boldsymbol{w}^{(\mu)})^T \in \mathbb{Z}_N^{n-1}$ is inherently independent of $\mu \in \{0, 1\}$. Therefore, our focus should primarily be on the last row of $\boldsymbol{Y} \cdot (\boldsymbol{w}^{(\mu)})^T \in \mathbb{Z}_N^n$, which can be expressed as

$$Y_{top} \cdot (\boldsymbol{w}^{(\mu)})^T \bmod N = (\langle \boldsymbol{x}, \boldsymbol{x}^{(\mu)} \rangle + a_\eta \cdot \langle \boldsymbol{x}, \boldsymbol{s} \rangle) \bmod N \tag{14}$$

We make the assumption that $\mathcal{A}$ can deduce that the information pertaining to $\boldsymbol{s} = (s_1, ..., s_n)$ is entirely ascertained by $Y_{top} \cdot \boldsymbol{s}^T \in \mathbb{Z}^{n-1}$. Consider $\boldsymbol{s}^{(0)} = (s_1^{(0)}, ... s_n^{(0)}) \in \mathbb{Z}^n$ as an random vector that satisfies $Y_{top} \cdot \boldsymbol{s}^T = Y_{top} \cdot (\boldsymbol{s}^{(0)})^T$ and $h_j = g^{s_j^{(0)}} \bmod N^2$ for all $j \in \{1, ..., n\}$. From the perspective of $\mathcal{A}$, the master key $\boldsymbol{s} \in \mathbb{Z}^n$ follows a distribution represented as $\boldsymbol{s}^{(0)} + \mathcal{D}_{\Lambda, \delta, -\boldsymbol{s}^{(0)}}$, where

$$\Omega = \{\boldsymbol{v} \in \mathbb{Z}^n | Y_{top} \cdot \boldsymbol{v} = 0 \text{ and } \boldsymbol{v} = 0 \bmod p'q'\} \tag{15}$$

Obviously, Ω constitutes the lattice $(p'q') \cdot \mathbb{Z} \cdot \boldsymbol{x}$. Conditionally on $\{h_j\}_{j=1}^n$ ans $Y_{top} \cdot \boldsymbol{s}$, the distribution of $\langle \boldsymbol{s}, \boldsymbol{x} \rangle$ is thus $\langle \boldsymbol{s}^{(0)}, \boldsymbol{x} \rangle + \mathcal{D}_{(p'q')||\boldsymbol{x}||^2 \cdot \mathbb{Z}, ||\boldsymbol{x}||\delta, -c}$, where $c = \langle \boldsymbol{s}^{(0)}, \boldsymbol{y} \rangle \in \mathbb{Z}$. Drawing upon the Lemma 8 from the comprehensive version of [7], given that $|(p'q')||\boldsymbol{x}||^2 \cdot \mathbb{Z}| < N^2||\boldsymbol{x}||^2$, $gcd(p'q'||\boldsymbol{x}||^2, N) = 1$ and $||\boldsymbol{x}||^2 < N$, selecting $\sigma > \sqrt{\lambda} \cdot N^{\frac{5}{2}}$ is sufficient to guarantee that $\langle \boldsymbol{s}, \boldsymbol{x} \rangle \bmod N$ deviates from the uniform distribution by no more than $2^{-\lambda}$ over $((p'q')||\boldsymbol{x}||^2 \cdot \mathbb{Z})/((p'q')||\boldsymbol{x}||^2 \cdot (N\mathbb{Z})) \cong \mathbb{Z}_N$. Since a_η is primarily invertible in $\mathbb{Z}_N$ with all but insignificant probability, the right side of Eq. (14) conceals the term $\langle \boldsymbol{x}, \boldsymbol{x}^{(\mu)} \rangle \bmod N$ statistically. When assessing probabilities, we can conclude that the advantage of $\mathcal{A}$ in the actual game could be confined to

$$|Pr[S_0] - \frac{1}{2}| \leq Adv_{\mathcal{A}}^{DCR}(\lambda) + \frac{1}{2^{\lambda-1}} \tag{16}$$

Therefore, it can be considered negligible if the DCR assumption holds.

5.3   The Master Secret Key Hiding of IBVE-IPE Scheme

In the IBVE-IPE scheme, the master secret key includes a vector $\boldsymbol{s}$, an element t and a bound parameter Y. We focus on analyze the hiding of the vector $\boldsymbol{s}$ here. Considering the linearity of the inner product, it is possible to recover the vector $\boldsymbol{s}$ by solving a system of linear equations in some previous works. The property necessitates that $\boldsymbol{s}$ remains concealing at all time. The IBVE-IPE performs this property, and we keep the subsequent theorem.

Theorem 3. *The IBVE-IPE scheme is master secret key hiding.*

Proof: For establishing the theorem, we make the assumption that $\mathcal{A}$ has control over multi decryptors to get the master secret key. The specifics are outlined below.

- **Setup**: The challenger $\mathcal{C}$ produces the key pair by executing **Setup**$(1^\lambda, 1^n, X, Y)$, which includes $mpk = (g, N, H, h, \{h_i\}_{i=1}^n, X)$ and $msk = (t, \boldsymbol{s}, Y)$. Subsequently, it transmits mpk to $\mathcal{A}$.
- **Query**: $\mathcal{A}$ randomly selects vectors $\boldsymbol{y}^{(1)}, ..., \boldsymbol{y}^{(l)}$ and ID^* to initiate queries to $\mathcal{C}$. $\mathcal{C}$ executes **KeyGen**$(msk, ID^*, \boldsymbol{y}^{(i)})$ to produce $sk_1^{(i)} = \langle \boldsymbol{s}, \boldsymbol{y}^{(i)} \rangle + t\alpha^{(i)} + H(ID^*)$, $sk_2^{(i)} = \alpha^{(i)} + t^{-1}H(ID^*)$, and $pk_{\boldsymbol{y}^{(i)}} = h^{\alpha^{(i)}} \bmod N^2$ where $\alpha^{(i)} \in_R \mathbb{Z}_N^*$. It then transmits the subsequent elements to $\mathcal{A}$, $\{sk_{\boldsymbol{y}^{(i)}} = (sk_1^{(i)}, sk_2^{(i)})\}_{i \in [1,l]}$ and $\{pk_{\boldsymbol{y}^{(i)}}\}_{i \in [1,l]}$.

– **Guess**: $\mathcal{A}$ provides a guess for vector $\boldsymbol{s}^* = (s_1^*, ..., s_n^*)$ and t^*. $\mathcal{A}$ can get information about the vector $\boldsymbol{s}^*$ and t^* in two manners.

Manner 1: Brute Force Hacking. $\mathcal{A}$ selects a random element $v \in \mathcal{D}_{\mathbb{Z}^n,\sigma}$ to calculate g^v, which is used to match the elements within $\{h_i\}_{i\in[1,n]}$ and determine if v will correspond to s_i within it. This is a form of brute force hacking, which is generally a strategy used to attack schemes that are not feasible.

Manner 2: $\mathcal{A}$ can utilize the requested keys $\{sk_{y^{(i)}}\}_{i\in[1,n]}$ to form a set of linear equations in this manner, particularly when $l \geq n$, in an attempt to deduce s by solving this set of linear equations. It's worth noting that among these $\boldsymbol{y}^{(1)}, ..., \boldsymbol{y}^{(l)}$, they are not linearly correlated.

$$\begin{cases} \langle \boldsymbol{s}, \boldsymbol{y}^{(1)} \rangle + t \cdot \alpha^{(1)} + H(ID^*) = sk_1^{(1)} \\ \quad \cdots \\ \langle \boldsymbol{s}, \boldsymbol{y}^{(l)} \rangle + t \cdot \alpha^{(l)} + H(ID^*) = sk_1^{(l)} \end{cases} \quad (17)$$

Next, we examine the possibility of success using this approach.

For retrieving $\boldsymbol{s}$ and t through manner 2, $\mathcal{A}$ would need to ascertain the random elements $t \in \mathbb{Z}_N$ and $\alpha^{(i)} \in \mathbb{Z}_N$ for all $i \in [1, l]$. We label the event where $\mathcal{A}$ successfully determines t and $\alpha^{(i)} \in \mathbb{Z}_N$ as Λ and Γ_i, respectively. Due to the inherent randomness associated with t and $\alpha^{(i)}$. It is reasonable to estimate that $Pr[\Lambda] = Pr[\Gamma_i] = \frac{1}{N}$. Therefore, we can obtain

$$Pr[(t = t^*) \wedge (\boldsymbol{s} = \boldsymbol{s}^*)] = Pr[\Lambda \wedge \Gamma_1 \wedge ... \wedge \Gamma_l]$$
$$= Pr[\Lambda] \cdot \prod_{i=1}^{l} Pr[\Gamma_i] \quad (18)$$
$$= \left(\frac{1}{N}\right)^{l+1}$$

As each event $\alpha^{(i)}$ of choice is independent of the others, this suggests that $\mathcal{A}$ can accurately obtain $\boldsymbol{s}$ and t with a negligible probability of $\left(\frac{1}{N}\right)^{l+1}$. Hence, the proposed scheme effectively conceals the master secret key.

5.4 Resistant Key Modify Attack of IBVE-IPE Scheme

To prove that our IBVE-IPE scheme can resists key modify attacks, we present the ensuing theorem.

Theorem 4. *The IBVE-IPE scheme is resists key modify attack.*

Proof. To prove this theorem, we simulate the real situation by the interaction between $\mathcal{A}$ and $\mathcal{C}$. The following is the specific process.

– **Init**: $\mathcal{A}$ selects and publicly discloses the challenge identity ID^* along with the vector $\boldsymbol{y}^*$.

- **Setup**: The challenger $\mathcal{C}$ produces the key pair by running $\mathbf{Setup}(1^\lambda, 1^n, X, Y)$, which includes $mpk = (g, N, H, h, \{h_i\}_{i=1}^n, X)$ and $msk = (t, \boldsymbol{s}, Y)$. It then sends mpk to $\mathcal{A}$.

- **Query**: $\mathcal{A}$ is permitted to request $\mathcal{C}$ for l keys $sk_{y^i}^{(ID^i)} (i \in [1, l])$, where the constraint is $ID^i \neq ID^j (i, j \in [1, l])$, and the challenge identity ID^* must be bound to the challenge vector $\boldsymbol{y}^*$. When $\mathcal{A}$ queries for ID^i and $\boldsymbol{y}^i$, $\mathcal{C}$ executes $\mathbf{KeyGen}(msk, ID^*, \boldsymbol{y}^{(i)})$ to produce $sk_1^{(i)} = \langle \boldsymbol{s}, \boldsymbol{y}^{(i)} \rangle + t\alpha^{(i)} + H(ID^*)$, $sk_2^{(i)} = \alpha^{(i)} + t^{-1}H(ID^*)$, and $pk_{y^{(i)}} = h^{\alpha^{(i)}} \bmod N^2$ where $\alpha^{(i)} \in_R \mathbb{Z}_N^*$. Afterward, it transmits the subsequent components to $\mathcal{A}$: $\{sk_{y^{(i)}} = (sk_1^{(i)}, sk_2^{(i)})\}_{i \in [1, l]}$ and $\{pk_{y^{(i)}}\}_{i \in [1, l]}$. When $\mathcal{A}$ queries for ID^* and $\boldsymbol{y}^*$, then $\mathcal{C}$ returns $\{sk_{y^{(*)}}^{(ID^*)} = (sk_1^{(*)}, sk_2^{(*)})\}_{i \in [1, l]}$ and $\{pk_{y^{(*)}}^{(ID^*)}\}_{i \in [1, l]}$ to $\mathcal{A}$.

- **Forge**: $\mathcal{A}$ has l keys $sk_{y^{(i)}}^{(ID^i)}$ and $sk_{y^{(*)}}^{(ID^*)}$. $\mathcal{A}$ needs to submit a forged key $sk_{y^{(')}}^{(ID^*)}$, where $\boldsymbol{y}' \neq \boldsymbol{y}^*$. $\mathcal{A}$ has the following information.

$$\begin{cases} sk_1^{(1)} = \langle \boldsymbol{s}, \boldsymbol{y}^{(1)} \rangle + t\alpha^{(1)} + H(ID^1) \\ \quad \cdots \\ sk_1^{(l)} = \langle \boldsymbol{s}, \boldsymbol{y}^{(l)} \rangle + t\alpha^{(l)} + H(ID^l) \\ sk_1^{(*)} = \langle \boldsymbol{s}, \boldsymbol{y}^{(*)} \rangle + t\alpha^{(*)} + H(ID^*) \end{cases} \tag{19}$$

$$\begin{cases} sk_2^{(1)} = \alpha^{(1)} + t^{-1}H(ID^1) \\ \quad \cdots \\ sk_2^{(l)} = \alpha^{(l)} + t^{-1}H(ID^l) \\ sk_2^{(*)} = \alpha^{(*)} + t^{-1}H(ID^*) \end{cases} \tag{20}$$

From this $\mathcal{A}$ tried to forge $sk_1' = \langle \boldsymbol{s}, \boldsymbol{y}' \rangle + t\alpha' + H(ID^*)$ and $sk_2' = \alpha' + t^{-1}H(ID^*)$.

$\mathcal{A}$ can assume that $\boldsymbol{y}' = k_1\boldsymbol{y}^1 + k_2\boldsymbol{y}^2 + ... + k_l\boldsymbol{y}^l$, where $k_1, k_2, ..., k_l$ are positive integers. $\mathcal{A}$ can compute $sk_1'' = k_1 sk_1^{(1)} + k_2 sk_1^{(2)} + ... + k_l sk_1^{(l)}$. Because of the randomness of $\alpha^{(i)}$, $\mathcal{A}$ can assume that $\alpha' = k_1\alpha^{(1)} + ... + k_l\alpha^{(l)}$. So $sk_1'' = \langle \boldsymbol{s}, \boldsymbol{y}' \rangle + t\alpha' + k_1 H(ID^1) + ... + k_l H(ID^l)$. $\mathcal{A}$ can let $sk_2'' = k_1 sk_2^{(1)} + ... + k_l sk_2^{(l)} = \alpha' + t^{-1}(k_1 H(ID^1) + ... + k_l H(ID^l))$. However, $\mathcal{A}$ needs to replace $k_1 H(ID^1) + ... + k_l H(ID^l)$ with $H(ID^*)$ in sk_1'' and $k_1 H(ID^1) + ... + k_l H(ID^l)$ with $H(ID^*)$ in sk_2'' to complete the key forgery. Since $\mathcal{A}$ does not know the t, and it cannot extract the value of $s^{-1}(k_1 H(ID^1) + ... + k_l H(ID^l))$ from sk_2'' alone. Therefore it cannot to forge the key by performing calculations, only by random guesses. So $\mathcal{A}$ forges the key successfully with negligible probability advantage $Adv_{IKA} = (\frac{1}{N})^{1+l}$. Therefore, our IBVE-IPE scheme can resists key modify attacks.

Table 1. Functional Comparison of Our Scheme With Existing Works

Work	Discrete Logarithm	Master Key Hiding	Key Verification	Identity Specific
[3]	Yes/No	No	No	No
[4]	Yes/No/No	No	No	No
[7]	Yes/No/No	No	No	No
[18]	No	Yes	No	No
Ours	No	Yes	Yes	Yes

Table 2. Computational Overhead of Our Scheme Compared to the Scheme in [18]

Algorithm	Yang *et al.* [18]	Our Scheme
Setup	$(n+1)E + (n+3)R + M$	$(n+2)E + (n+4)R + M$
KeyGen	$2E + 2R + nM$	$E + (n+2)M + R + I + H$
Encrypt	$(n+4)E + (3n+1)M + 4R$	$(n+3)E + (3n+1)M + R + H$
Verify	—	$(n+2)E + M$
Decrypt	$(n+4)(E+M) + 2I$	$(n+3)(E+M) + 2I + H$

6   Performance Evaluation

We perform the performance analysis of the presented scheme in this section, which consists of functionality analysis and efficiency analysis.

6.1   Functionality Analysis

In Table 1, we present a functional comparison between our proposed IBVE-IPE scheme and several existing schemes [3,4,7], and [18]. This comparison is the requirement to calculate discrete logarithms, master key hiding, key verification and identity specification, respectively. From Table 1, it becomes evident that the scheme presented in this paper holds greater application potential. The IPFE scheme relying on the Decisional Diffie-Hellman (DDH) assumption, as presented in [3], serves as the pioneering instance of the innovative IPFE concept. However, this scheme necessitates a high overhead computation of discrete logarithms in the decryption algorithm to gain the required inner product. Consequently, this instantiation scheme, which is based on ElGamal encryption, is confined to computing inner product values within a limited range, making it impractical in terms of computational overhead. To overcome this limitation, subsequent schemes such as [4,7] based on LWE and schemes [7,18] based on the DCR assumption have been proposed. In these different schemes, the computation of discrete logarithm can be circumvented, allowing for the retrieval of the required inner product without constraints on its range. Regarding the issue of the privacy of the key vector consist of n components, previous research [7] explored this challenge within a subspace formed by n-1 linearly independent vectors. However, the aspect of ensuring complete master secret key concealment across the entire space is not thoroughly examined. In contrast, our scheme and the work presented in [18] both address the issue of concealing the master secret key across the entirety of the space. Additionally, our scheme enables the specification of the recipient's identity and key verification.

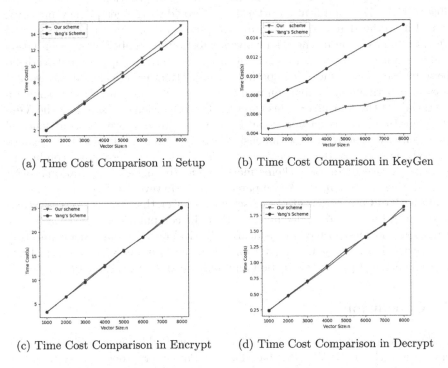

(a) Time Cost Comparison in Setup   (b) Time Cost Comparison in KeyGen

(c) Time Cost Comparison in Encrypt   (d) Time Cost Comparison in Decrypt

Fig. 1. Time Cost Comparisons for Four Algorithms

6.2 Efficiency Analysis

Table 2 offers a comprehensive comparison of the computational overhead between our scheme and the work presented in [18]. Our objective is to demonstrate that we achieve superior privacy preserving and offer more features while maintaining reasonable computational overhead. Since both schemes are based on DCR assumptions, comparing them is a reasonable approach. Within this table, we represent an exponentiation operation, the selection of a random value, a multiplication operation, a hash operation and inversion operation as E, R, M, H, I, respectively.

As shown in Table 2, during the **Setup** phase, both schemes require similar operations for system initialization, including exponentiation operation, multiplication operation and the selection of a random value. However, because our scheme includes additional parameters, there is a slight increase in overhead. Moving on to the **KeyGen** phase, it's worth noting that the computational overhead of exponentiation operation significantly contributes to the overall cost. In this regard, our scheme exhibits lower computational overhead compared to the scheme in [18]. Since our scheme introduces an additional **Verify** algorithm, although it requires some exponentiation and multiplication operations, it is able to resist key modification attacks and provides stronger privacy preservation. Its overhead remains within an acceptable range. For the **Encrypt** and **Decrypt** phases, both schemes involve a substantial number of exponentiation and multiplication operations. Consequently, the overall computational overhead between the two schemes is comparable. Besides the theoretical analysis, we also present experimental comparison results in Fig. 1. We compare the experimental overheads of the four algorithms common to both schemes, excluding the **Verify** algorithm as it is an additional component. To improve the confidence of the results, we control vector dimensions ranging from 1000 to 8000 for experimental comparisons. It is evident that under the same experimental conditions, as depicted in Fig. 1a, our scheme incurs a slightly higher computational overhead during the **Setup** phase compared to [18]. However, in Fig. 1b, we observe that the **KeyGen** overhead of our scheme is lower than that of [18]. Furthermore, both Fig. 1c and 1d demonstrate that the computational overhead for the **Encrypt** and **Decrypt** phases is nearly identical between our scheme and [18]. As previously mentioned, any additional time costs in our scheme are justified by the enhanced privacy protection and additional features it offers.

7   Conclusion

In this paper, we present an identity-based key verifiable inner product function encryption scheme (IBVE-IPE) based on the Paillier cryptographic system. Our proposed IBVE-IPE scheme allows for the specification of the recipient's identity, hides the master secret key vector information, and is resistant to key modification attacks. This addresses the limitations of general inner product function encryption, which cannot specify the recipient, is vulnerable to key modification, and may leak master key vector information. While there may be sacrifices in computational efficiency, our scheme offers enhanced privacy and additional features. Furthermore, we provide the security analysis along with experimental findings that illustrate both the security and practicality of our presented scheme.

References

1. Abdalla, M., Benhamouda, F., Gay, R.: From single-input to multi-client inner-product functional encryption. In: Galbraith, S.D., Moriai, S. (eds.) ASIACRYPT 2019. LNCS, vol. 11923, pp. 552–582. Springer, Cham (2019). https://doi.org/10.1007/978-3-030-34618-8_19
2. Abdalla, M., Benhamouda, F., Kohlweiss, M., Waldner, H.: Decentralizing inner-product functional encryption. In: Lin, D., Sako, K. (eds.) PKC 2019. LNCS, vol. 11443, pp. 128–157. Springer, Cham (2019). https://doi.org/10.1007/978-3-030-17259-6_5
3. Abdalla, M., Bourse, F., De Caro, A., Pointcheval, D.: Simple functional encryption schemes for inner products. In: Katz, J. (ed.) PKC 2015. LNCS, vol. 9020, pp. 733–751. Springer, Heidelberg (2015). https://doi.org/10.1007/978-3-662-46447-2_33
4. Abdalla, M., Bourse, F., Caro, A.D., Pointcheval, D.: Better security for functional encryption for inner product evaluations. Cryptology ePrint Archive, Paper 2016/011 (2016). https://eprint.iacr.org/2016/011
5. Abdalla, M., Bourse, F., Marival, H., Pointcheval, D., Soleimanian, A., Waldner, H.: Multi-client inner-product functional encryption in the random-oracle model. In: Galdi, C., Kolesnikov, V. (eds.) SCN 2020. LNCS, vol. 12238, pp. 525–545. Springer, Cham (2020). https://doi.org/10.1007/978-3-030-57990-6_26
6. Abdalla, M., Catalano, D., Fiore, D., Gay, R., Ursu, B.: Multi-input functional encryption for inner products: function-hiding realizations and constructions without pairings. In: Shacham, H., Boldyreva, A. (eds.) CRYPTO 2018. LNCS, vol. 10991, pp. 597–627. Springer, Cham (2018). https://doi.org/10.1007/978-3-319-96884-1_20
7. Agrawal, S., Libert, B., Stehlé, D.: Fully secure functional encryption for inner products, from standard assumptions. In: Robshaw, M., Katz, J. (eds.) CRYPTO 2016. LNCS, vol. 9816, pp. 333–362. Springer, Heidelberg (2016). https://doi.org/10.1007/978-3-662-53015-3_12
8. Asharov, G., Segev, G.: Limits on the power of indistinguishability obfuscation and functional encryption. SIAM J. Comput. **45**(6), 2117–2176 (2016)
9. Boneh, D., Sahai, A., Waters, B.: Functional encryption: definitions and challenges. In: Ishai, Y. (ed.) TCC 2011. LNCS, vol. 6597, pp. 253–273. Springer, Heidelberg (2011). https://doi.org/10.1007/978-3-642-19571-6_16
10. Chotard, J., Dufour Sans, E., Gay, R., Phan, D.H., Pointcheval, D.: Decentralized multi-client functional encryption for inner product. In: Peyrin, T., Galbraith, S. (eds.) ASIACRYPT 2018. LNCS, vol. 11273, pp. 703–732. Springer, Cham (2018). https://doi.org/10.1007/978-3-030-03329-3_24
11. Goldwasser, S., et al.: Multi-input functional encryption. In: Nguyen, P.Q., Oswald, E. (eds.) EUROCRYPT 2014. LNCS, vol. 8441, pp. 578–602. Springer, Heidelberg (2014). https://doi.org/10.1007/978-3-642-55220-5_32
12. Paillier, P.: Public-Key cryptosystems based on composite degree residuosity classes. In: Stern, J. (ed.) EUROCRYPT 1999. LNCS, vol. 1592, pp. 223–238. Springer, Heidelberg (1999). https://doi.org/10.1007/3-540-48910-X_16
13. Regev, O.: On lattices, learning with errors, random linear codes, and cryptography. J. ACM **56**(6), 1–40 (2009)
14. Rivest, R.L., Shamir, A., Adleman, L.: A method for obtaining digital signatures and public-key cryptosystems. Commun. ACM **21**(2), 120–126 (1978)
15. Shi, E., Chan, H., Rieffel, E., Chow, R., Song, D.: Privacy-preserving aggregation of time-series data. ACM Trans. Sen. Netw **5**(3), 1–36 (2009)

16. Waters, B.: A punctured programming approach to adaptively secure functional encryption. In: Gennaro, R., Robshaw, M. (eds.) CRYPTO 2015. LNCS, vol. 9216, pp. 678–697. Springer, Heidelberg (2015). https://doi.org/10.1007/978-3-662-48000-7_33

17. Wong, W.K., Cheung, D.W.l., Kao, B., Mamoulis, N.: Secure kNN computation on encrypted databases. In: Proceedings of the 2009 ACM SIGMOD International Conference on Management of data, pp. 139–152 (2009)

18. Yang, H., Su, Y., Qin, J., Wang, H.: Privacy-preserving outsourced inner product computation on encrypted database. IEEE Trans. Dependable Secure Comput. **19**(2), 1320–1337 (2020)

19. Zhang, M., Huang, S., Shen, G., Wang, Y.: PPNNP: a privacy-preserving neural network prediction with separated data providers using multi-client inner-product encryption. Comput. Stand. Interfaces **84**, 103678 (2023)

20. Zhang, M., Li, Z.A., Zhang, P.: A secure and privacy-preserving word vector training scheme based on functional encryption with inner-product predicates. Comput. Stand. Interfaces **86**, 103734 (2023)

DRSA: Debug Register-Based Self-relocating Attack Against Software-Based Remote Authentication

Zheng Zhang[1], Jingfeng Xue[1], Tianshi Mu[2], Ting Yu[2], Kefan Qiu[1], Tian Chen[1], and Yuanzhang Li[1(✉)]

[1] Beijing Institute of Technology, Beijing 100081, China
popular@bit.edu.cn
[2] China Southern Power Grid Digital Grid Group Co., Ltd., Guangzhou 510000, Guangdong, China

Abstract. Remote attestation (RA) is an essential feature in many security protocols to verify the memory integrity of remote embedded (IoT) devices. Several RA techniques have been proposed to verify the remote device binary at the time when a checksum function is executed over a specific memory region. A self-relocating malware may try to move itself to avoid being "caught" by the checksum function because the attestation provides no information about the device binary before the current checksum function execution or between consecutive checksum function executions. Several software-based that lack of dedicated hardware rely on detecting the extra latency incurred by the moving process of self-relocating malware by setting tight time constraints. In this paper, we demonstrate the shortcomings of existing software-based approaches by presenting Debug Register-based Self-relocating Attack (DRSA). DRSA monitors the execution of the checksum function using the debug registers and erases itself before the next attestation. Our evaluation demonstrates that DRSA incurs low overhead, and it is extremely difficult for the verifier to detect it.

Keywords: Remote attestation · Debug registers · Self-relocating malware

1 Introduction

With the development of the Internet of Things and the increase in the variety and number of special-purpose embedded devices, the security threats faced by these devices are increasing. Preferred targets of malware change from general-purpose computers to numerous and inter-connected embedded devices mainly because of the latter's lack of security protections. As our society is gradually becoming surrounded by these low-end embedded devices, their safety cannot be ignored, although security is typically not the highest priority due to cost, power constraints, or size.

© ICST Institute for Computer Sciences, Social Informatics and Telecommunications Engineering 2024
Published by Springer Nature Switzerland AG 2024. All Rights Reserved
J. Chen and Z. Xia (Eds.), BlockTEA 2023, LNICST 577, pp. 23–40, 2024.
https://doi.org/10.1007/978-3-031-60037-1_2

For such devices that have no means to prevent malware attacks, Remote Attestation (RA) is a distinct security service for detecting malware on embedded devices. RA is typically realized as a challenge-response protocol. In this protocol, RA allows a trusted Verifier ($\mathcal{VRF}$) to compute a checksum of an untrusted Prover ($\mathcal{PRV}$)'s memory to attest its integrity. Since the $\mathcal{VRF}$ is assumed to know the hardware configuration and exact memory contents of the $\mathcal{PRV}$, it can compute the expected checksum and compare it with the received one. Mismatched values indicate that the device has most likely been compromised. Software-based RA [16–18,20] is an approach particularly suitable for verifying the trustworthiness of low-end embedded devices because it does not rely on hardware to control the execution of the integrity-ensuring function. Compared with hybrid RA (based on hardware/software co-design) [1,6,10] and hardware-based RA (e.g., those based on a TPM [11] or other dedicated hardware modules), which require some level of hardware support, software-based RA is less costly.

Most of the previously proposed software-based RA techniques perform an integrity check on the whole memory content of a low-end sensor node because the memory of this kind of device is usually small enough. Due to the rapid technological growth, the usage of middle-end and high-end devices is increasing. These devices can perform tentative computations such as deep learning network training [13,22], executing complex protocols and authentication algorithms [23, 24]. In middle-end and high-end devices that have enough resources, such as a powerful processor with x86 architecture, to run a traditional OS such as Linux, however, attesting the whole memory content is time-consuming.

In this paper, we emphasize that the security of software-based attestation on general-purpose operating systems is uncertain. We present a Debug Register-based Self-relocating Attack (DRSA), a malicious program that circumvents attestation by relocating itself and restoring the contents before it is measured. The implementation of this attack uses debug registers provided by the debugging architecture, which is available on most processors today. In general, debug registers can support software/hardware breakpoints, and the trigger conditions are various. DRSA utilizes debug registers to monitor the beginning and the end of an RA measurement, erases itself, and restores the original program memory contents before attestation. The malicious code can be restored after the RA procedure to regain control over the compromised device. We evaluate DRSA and show the shortcomings of software-based attestation design.

The remainder of this paper is organized as follows: Sect. 2 provides a preliminary on software-based attestation technologies and the x86 debugging architecture. Section 3 discusses the threat model and assumptions of this work. Section 4 presents a detailed description of DRSA. Section 5 introduces the design and implementation details. The effectiveness and efficiency of DRSA are evaluated and discussed in Sect. 6. Section 7 discusses related work and Sect. 8 concludes this paper.

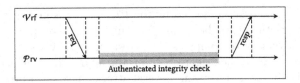

Fig. 1. Timeline of a typical RA protocol

2 Background

2.1 Software-Based Attestation

Existing software-based RA techniques are all realized as a challenge-response protocol where the $\mathcal{VRF}$ challenges a $\mathcal{PRV}$ (a target untrusted remote device) to compute a checksum of its specified memory region. We first describe the basic challenge-response protocol and then list the existing software-based RA schemes.

Challenge-Response Protocol. A typical challenge-response paradigm uses a checksum function to compute the checksum of the memory content, the procedure of which can be described as follows:

(1) $\mathcal{VRF}$ sends an attestation request with a challenge to $\mathcal{PRV}$. This request also contains a nonce generated by the $\mathcal{VRF}$ to be involved in the checksum computing to prevent pre-computation or replay attacks.
(2) $\mathcal{PRV}$ receives the request and computes a challenge-based authenticated integrity check over a pre-defined memory region. Depending on the attestation purpose, this region corresponds to either the entire program memory to ensure that no malware exists or only a specific memory region to obtain the guarantee of a dynamic root on the untrusted platform [16].
(3) $\mathcal{PRV}$ returns the result to $\mathcal{VRF}$.
(4) $\mathcal{VRF}$ receives the result and checks whether it matches the expected checksum.

The timeline illustrating this attestation process is shown in Fig. 1. Since the usual RA threat model assumes that the target device is fully compromised, software-based RA must protect the integrity-ensuring function from the potential malware without any hardware support. Prior works rely on strict time constraints or filling the empty memory locations to prevent malicious code.

Time Constraint. A compromised $\mathcal{PRV}$ may try to evade the detection of software-based RA schemes by performing a memory copy attack. This attack modifies the original code and redirects the memory accesses to a correct copy of the original code that is stored elsewhere in memory. Some software-based RA relies on a strict time constraint and optimized attestation code to defend against memory copy attack because they claim that the overhead caused by redirection would be easily detected by the $\mathcal{VRF}$. Checksum functions of these approaches must, therefore, be simple and highly time-optimized, leaving no room for the adversary to compress the attest time. However, it is difficult to assess what the best attack against a certain approach is, which makes the security of this approach uncertain. Several attacks have been proposed [7] and demonstrated the weaknesses of time constraints. Moreover, most time-based attestation require disabling interrupts during attestation procedure execution to ensure that the adversary can not move malware around during attestation by forcing an interrupt. This will affect the normal operation of the device and result in low availability.

Memory Filling. To overcome the weakness of tight time constraints, some software-based RA schemes fill the free program memory with random noise before deployment so that there is no empty space for the adversary to store its malware. These approaches claim that the attacker can not simply overwrite these noises because they are involved in the calculation of the checksum. However, this method still has disadvantages. In fact, the adversary can compress the original code to gain free space for storing its malware and executing it. Moreover, memory filling is only suitable for sensor devices where the memory layout is known in advance. For some implementations based on complex instruction sets and operating systems, the situation will be more complicated. For example, Pioneer is a software-based RA implemented on x86 architecture with Linux kernel, the memory layout of which may change dynamically during program execution due to the dynamic memory allocation. Memory filling will not work because the pre-filled noises may be overwritten during program execution.

2.2   X86 Debugging Architecture

The debugging architecture is supported by most of the processors today to facilitate on-chip debugging. There are a total of eight debug registers (DR0 through DR7) in x86 architecture to support both debug exceptions and hardware breakpoints required by software debuggers [12]. Among them, DR0 to DR3 are used to specify memory addresses or I/O locations to be monitored by the debugger. The processor raises a debug exception when an instruction or memory address matches the address that is stored in one of these four registers. The address monitoring process is performed in parallel with the normal virtual to physical address translation, and thus no additional expensive intercept-and-check in software is required. DR4 and DR5 are reserved, while DR6 is used to report the status of the debugging conditions when a debugging exception is raised. Bn bit of DR6 indicates that the nth breakpoint was reached. BS and BT bits signify the exception is triggered by single stepping and task switching, respectively. DR7 is used to control/configure the trigger condition of the debug exception when the address matches. By setting the corresponding Ln or Gn bit in DR7, the nth breakpoint could be enabled or disabled. Ln bit and Gn bit enable the nth breakpoint for the current task and all tasks, respectively. The corresponding breakpoint is disabled when both Ln and Gn bits are cleared. R/Wn field configures the trigger mode of the nth breakpoint. For example, the data read or write in the monitored memory address will raise a debugging exception if value 11 is set. The instruction execution of the specific memory address can also be realized by monitoring the set value 00. The size of the monitored memory location of the nth breakpoint can be set through the LENn field.

3 Threat Model and Assumptions

3.1 Hardware Platform Description

DRSA is implemented on devices with x86 architecture. Unlike the Harvard memory architecture, the program and data memories of x86 architecture devices are not physically separated. This makes self-relocating malware more difficult to evade attestation on x86 architecture. In fact, the separation of program memory and data memory gives the attacker a vulnerability to evade attestation. The attacker can hide malicious code in data memory and pass the attestation protocol that does not attest data memory. While the presented attack is validated on an x86 architecture-based device, it is not specific to x86 architecture. It exploits the characteristics of the debugging architecture supported by the processor. The proposed attack is applicable to any device that uses similar debugging features, such as tracing and monitoring.

3.2 Adversarial Assumptions

In this work, we assume that the adversary aims to install its malicious code in the memory of the target device and avoid being detected during the attestation process. The adversary has complete control over the code and data of the target device before and after attestation. We also assume that the attacker has no direct control of the device at attestation time. The attack succeeds if the device passes the attestation while the malicious code resides in memory. We do not address the details of how attackers install malicious code on devices because it is beyond the scope of this paper. Finally, we assume that the attacker does not perform hardware attacks on the target device. Specifically, it does not modify the device hardware or induce hardware faults.

4 Debug Register-Based Self-relocating Attack

In this section, we introduce the implementation of DRSA. The attack is formed based on the debugging features of x86 debugging architecture. It circumvents malware detection by detecting the start of the attestation and restoring the original contents before it is measured. This is achieved by monitoring instruction execution and data access events using debug registers.

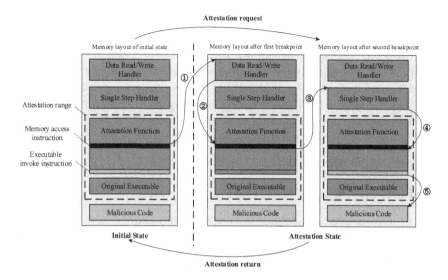

Fig. 2. Overview of DRSA. The numbers represent the transfer order of the control flow.

Figure 2 illustrates the overview of DRSA. The memory layout of the target device is also depicted. The core of the attack is composed of a data read and write handler, a single-step handler, a hook (a jump instruction), and malicious code. When an attestation request is received, the target device switches from the initial state to the attestation state, and its memory layout also changes. The numbers represent the transfer order of the control flow. As stated above, the memory layout of the target device with x86 architecture is not divided into program memory and data memory. Moreover, the existing remote attestation routine for x86 architecture, Pioneer, only verifies part of the memory content rather than the entire memory [16]. The checksum code of Pioneer is self-checksumming and computes a checksum over the entire attestation function and the executable. The correct checksum provides a guarantee that the verification function code is unmodified and further establishes a trusted computing base called the dynamic root of trust. Once established, the dynamic root of trust verifies the integrity of the executable and invokes it. The executable is guaranteed to be trusted if it is unmodified and executed uninterrupted. DRSA works by inserting a hook (a jump instruction) into the attestation routine to replace the invoke instruction during the attestation procedure function. When the dynamic root of trust tries to invoke the origin executable, the control flow will be hijacked, and malicious code will be invoked.

However, simply modifying the attestation function will be detected during the attestation procedure because it will lead to an invalid checksum. To escape detection, DRSA utilizes two types of hardware breakpoints to monitor the checksum function's operation of reading the memory content and modifying the memory content in the corresponding exception handlers. Specifically, the following is a detailed description of the attack scenario presented in Fig. 1:

1. The malicious code is installed on the victim device beforehand, and the existing exception handlers are replaced with our own handlers to implement DRSA's core function. Next, we insert a hook to the attestation code by replacing the executable invoke instruction of the attestation function with a jump.

2. A data read/write breakpoint and a single-step breakpoint are set, respectively, to trigger the corresponding exception handler. Specifically, the data read/write breakpoint is set beforehand, while the single-step breakpoint is set at runtime.

3. When an attestation request is received, the victim device switches to the attestation state. The attestation function will access the memory of the attestation region and compute the checksum. The data read/write breakpoint will be triggered when the attestation function accesses the address of the hook, and the data read/write exception handler will be invoked.

4. The data read/write exception handler receives control and sets the original executable invoke instruction back to what it was. Then the single-step breakpoint is set before the control flow transfers back to execute the memory access instruction.

5. The single-step breakpoint is triggered after the single memory access instruction executes. The single-step exception handler receives control and installs the hook to replace the executable invoke instruction of the attestation function again. Then the control flow transfers back to continue executing the next instruction of the memory access instruction.

6. The attestation function will compute a valid checksum because the modified memory content is restored to the original content before being accessed by its memory access instruction. The victim device switches back to the initial state, but the hook, however, is still present in the memory without being detected by the verifier.

The attestation function is, therefore, executed over the clean memory of the attestation region. Since neither the exception handlers nor the malicious code are in the attestation region, their presence will not affect the checksum result. Thus, the valid checksum is sent to the verifier, and the integrity of the executable will be measured. Since the content of the executable has not been modified, it can pass the integrity measurement and is ready to be invoked. Once the original executable is invoked, the control flow will be redirected to execute the malicious code.

5   Design and Implementation

To achieve the attack process mentioned above, DRSA utilizes debugging architecture, which provides hardware breakpoints and exception mechanisms to transfer execution to custom handlers, which relocates malicious content before being detected. To achieve this, the corresponding bits of the debug registers should be set correctly. In this section, we explain the detailed setting of the debug registers and the key implementation steps of DRSA under the x86 architecture.

5.1   Memory Restoring via Data Read and Write Breakpoint

DRSA uses the debugging architecture of the underlying platform to provide hardware breakpoints on code execution, data, and memory access. Support for debugging architecture on most processors is in the form of debug registers. The breakpoint trigger conditions can have various attributes such as execution, write, read/write, etc. To achieve memory restoration via data read and write breakpoint, the address and attributes of the breakpoint should be set through corresponding registers.

In the x86 architecture, current processors typically provide 4 locations of hardware breakpoints, the addresses of which can be set through DR0-DR3. We use the first debug register, DR0, and set its value as the address of the hook to monitor any memory access to it. The specific attributes of the breakpoint are controlled by DR7. Table 1 shows the detailed setting of each bit of DR7. L0 bit and G0 bit control the scope of the first breakpoint. It can be set as persistent or nonpersistent during its creation by setting the G0 bit to 1 or setting the L0 bit to 1, respectively. We set the G0 bit to 1 so that every access to the hook address will trigger the read and write breakpoint. GD bit is set to prevent debug registers from being modified by other programs. Instructions that access debug registers will raise an exception, and the relevant information on debugging exceptions will be recorded by DR6. R/W0 is controlled by bits 16 and 17 of DR7, which determines the trigger condition of the breakpoint. We set R/W0 to 01 so that the breakpoint will be triggered by the read/write operation of the address set by DR0. Finally, LEN0 is set to 00 to specify the memory size monitored by the breakpoint as 1 byte.

Table 1. Dr7 settings for data read/write breakpoint.

Bit Name	Bit Number	value	Description
L0	0	1	the breakpoint will only be valid for the current task
G0	1	0	the breakpoint will be valid for all the tasks
GD	13	1	instructions that access debug registers will raise an exception
R/W0	16&17	01	the breakpoint will be triggered by read or write operations
LEN0	18&19	00	specify the memory size monitored by the breakpoint as 1 byte

DRSA installs its own data read and write handler, replacing the existing handler to implement the restoration of original memory content. When the data read and write handler is invoked due to a memory access exception, the handler first erases the hook code and then restores the original memory contents. Finally,

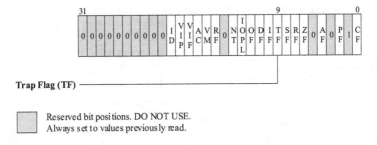

Fig. 3. The trap flag in the EFLAGS register

the handler prepares for a single-step exception and transfers the control flow back to the attestation function.

5.2 Hook Re-installing via Single-Stepping

DRSA employs a single-step exception for hook re-installing. The single-step exception is a standard exception on all processor architectures that are triggered upon execution of each instruction when a certain processor flag (also called the trap flag) is activated. The exception generated by a single step is consistent with the exception generated by the hardware breakpoint. Thus, the single-step exception makes use of the trap flag for its functioning. In the x86 architecture, the trap flag can be set through the EFLAGS register. The position of the trap flag in the EFLAGS register is shown in Fig. 3. DRSA installs its own single-step handler, replacing the existing handler to implement hook re-installing. The single-step handler is installed on demand from the data read and write handler.

In the data read and write handler, the trap flag (bit 8) is set to 1 to enable single-step mode for debugging before the control flow transfers back to execute the memory access instruction of the attestation function. Then, the single-step exception will be triggered after the execution of the memory access instruction. When the single-step handler is invoked due to a single-step exception, the handler reinstalls the hook back to replace the executable invoke instruction of the attestation function again. Then, the handler resets the trap flag, uninstalls itself, and issues an end of exception.

5.3 Framework Composition and Source Organization Under Windows

The framework of DRSA is implemented in the form of a loadable kernel mode driver under the Windows operating system that is able to monitor the memory access operation initiated by the attestation function and relocate malicious content before it is detected. Figure 4 illustrates how DRSA performs memory access monitor and malicious content relocating. The left part is the framework

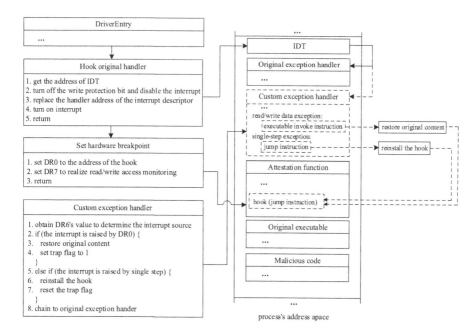

Fig. 4. DRSA in the form of a kernel mode driver

of the DRSA in the Windows driver form, while the right part is the attestation process's address space. The driver is mainly composed of three key parts (pseudo-code shown in Fig. 4):

Hook Original Handler: To install its own exception handlers, DRSA intercepts calls to the Interrupt Descriptor Table (IDT) and adds in its own processing. After the DriverEntry initializes driver-wide data structures and resources, the DRSA driver creates a hook to replace the original exception handler. Specifically, it first obtains the base address of the IDT and then extracts the address of the corresponding original handler function from an interrupt descriptor. The write protect bit is turned off (by modifying the CR0 register), and the interrupt is disabled (by using the CLI instruction) to avoid being interrupted by other interrupts when modifying the IDT. Finally, it points the target interrupt descriptor to a new handler function, reloading the CR0 register and enabling interrupts. When an exception is raised, the custom exception handler function will be invoked instead of the original exception handler.

Set Hardware Breakpoint: The data read/write breakpoint is set during the driver's loading process. It mainly includes setting the DR0 register, and the DR7 register. Specifically, the DR0 register is assigned the address of the hook we installed in the attestation function, while the DR7 register is assigned the specific value summarized in Sect. 5.1 so that the trigger condition of the breakpoint is set to read/write. During the driver development routine, we achieve direct control of the DR registers through inline assembly.

Custom Exception Handler: The custom exception handler is developed as part of the driver and loaded into memory when the driver is installed. To achieve malicious code relocating, the custom exception handler keeps a copy of the original memory content and malicious code to be installed in the attestation code. When it is invoked by a raised exception, the value of the DR6 register is first obtained to determine the interrupt source. The value of BD (bit 13), as mentioned in Sect. 5.1, indicates whether the exception is triggered by the breakpoint set by the DR0 register. The BD bit is 1, indicating that there is a program trying to access the memory address (read or write) monitored by the DR0 register. Then, the custom exception handler will restore the original content of the attestation code (an executable invoke instruction) and set the trap flag of the EFLAGS register to 1. Next, it will chain to the original exception handler to complete its original function. If the exception is not raised by the DR0 register, and the trap flag is set, which means that the exception is raised by single stepping, the custom exception handler will reinstall the hook and reset the trap flag and chain to the original exception handler. Other interrupt sources will not trigger the execution of additional code in the custom handler and will transfer the control flow to execute the original handler instead.

Owing to the capabilities that debug architecture provides, it is clear that the malicious modification of the target memory content can successfully escape the detection of the attestation function.

5.4 Framework Composition and Source Organization Under Linux

DRSA under Linux OS is implemented using ptrace, a system call provided by Unix, and several Unix-like operating systems. By attaching to another process using the ptrace call, a tool has extensive control over the operation of its target. It can also single-step through the target's code and observe and intercept system calls and their results. More importantly, ptrace provides requests to control the debug registers directly, which facilitates the implementation of our attack method. The framework of DRSA based on ptrace is shown in Fig. 5. DRSA is implemented as one single process (the "tracer") to observe and alter the execution of the attestation process (the "tracee"), achieving malicious code relocation at runtime.

The left part of Fig. 5 illustrates the execution flow of DRSA based on price. DRSA process commences tracing by having the attestation process do a *PTRACE_ATTACH*. Then, the attestation process will stop, and the DRSA process can use various ptrace requests to inspect and modify the attestation process. DRSA process first sets data read/write breakpoints at the hook address of the attestation process. By using ptrace requests *PTRACE_POKEUSER*, the DRSA process sets the DR0 register and DR7 register to the specific values summarized in Sect. 5.1. Then, the call *PTRACE_CONT* is delivered to continue the execution of the attestation process. The attestation process will stop again when the attestation function triggers the data read/write breakpoint. The DRSA process restores the original content of the attestation function by using *PTRACE_POKETEXT*, which can modify the code segment of the tracee

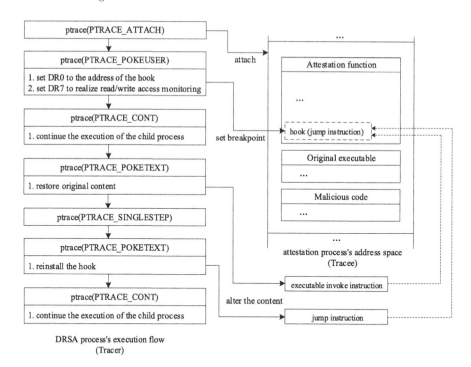

Fig. 5. Implementation of DRSA under Linux

directly. After that, the DRSA process issues a *PTRACE_SINGLESTEP* request to restart the stopped attestation process but arranges for it to be stopped after the execution of a single instruction (the memory access instruction in our scenario). The hook code (a jump instruction) will be reinstalled in the attestation function during this suspension, and the DRSA process then makes the place-stopped attestation process run again by using *PTRACE_CONT*.

By utilizing ptrace, the attestation process will not be aware of the existence of the DRSA process. Additionally, attaching to the tracee will not add additional instructions in the attestation region, which will not affect the result of the checksum calculation.

6   Evaluation

In this section, we measured the effectiveness of DRSA on some specific code attestation schemes. We tested our attack on SWATT [17] to compare with existing attack methods against the existing remote attestation routine. The untrusted platform used in our experiment is a PC with a 3.41 GHz Intel i7-6700 processor running Windows 10.

6.1   Attestation Time Overhead

In this subsection, we tested the time overhead of DRSA on existing protocols. Our goal is to show that the proposed attack could escape from the detection of the target attestation routine.

We tested DRSA on our implementation of SWATT. The original SWATT was implemented based on an ATMega163L microcontroller with 16 KBytes of program memory. In order to simulate SWATT on the x86 platform, we ported SWATT's attestation function and made it verify the specified 128 KBytes memory region, including itself. We further implemented an attacker's version that assumes that DRSA has compromised the attestation function of SWATT. Table 2 shows the results indicating the attack overhead introduced by DRSA. We compared it with the original SWATT attack. We also compared the attack overhead imposed by DRSA with those introduced by [7], including the memory shadowing attack and ROP attack. Note that we obtained the overhead of ROP attack measures from the articles previously published because it can only be implemented on a Harvard memory architecture.

Table 2. Overhead of different attacks.

Method	Time of Execution (ms)	Attack Overhead (ms)	Attack Overhead (%)
Original SWATT	11061	–	–
Original SWATT Attack	–	–	13%
SWATT (MicaZ)	13103	–	–
ROP Attack (MicaZ)	–	42.3	0.32%
SWATT (x86)	56.5	–	–
Shadow Attack (x86)	60.1	3.7	6.47%
DRSA	**56.9**	**0.4**	**0.71%**

To facilitate the comparison of different attack overhead, timings are collected on three implementation versions of SWATT. The original SWATT attack represents the best attack expected by the designers of SWATT, which checks each address generated in the pseudo-random sequence. This attack would add three cycles for test and redirection. Since the original attestation function main loop is 23 cycles long, the malicious attestation function incurs 13% attack overhead. The ROP attack is implemented on a MicaZ device with 128 KBytes of program memory in [7]. The authors ported SWATT on MicaZ and tested their ROP attack on it. The experimental results show that the time required for the ROP attack to hide the malicious code is less than 50 milliseconds, introducing about 0.32% overhead.

The last three rows in Table 2 summarize the execution time of x86-based SWATT, the attack overhead of DRSA, and the shadow attack based on our x86 platform. On the x86 platform, the number of SWATT cycles should be increased, according to the Coupon's Collector Problem. Due to the high CPU frequency of the experimental platform, we further double the number of iterations to get more accurate results. Compared to the running time of x86-based SWATT, DRSA introduced 0.71% of overhead. The experimental results show that the attack overhead of DRSA is much smaller than the original SWATT attack and shadow attack. While it is a bit slower than the ROP attack, this delay is hard to be detected by the verifier. The main reason for the difference in the overhead of DRSA and ROP attack is that the ROP attack inserts the hook and triggers the rootkit hiding functionality, which will delete the rootkit code from the program memory at the beginning of the attestation function. This means that additional malicious code is executed only once during an entire attestation routine. In DRSA, however, the execution times of the exception handlers depend on the number of times the modified memory location is accessed during an attestation routine. Therefore, additional malicious code may be triggered multiple times, which incurs more execution time.

6.2   Status Information on the Stack

The authors of [16] have taken into account that an adversary generates a nonmaskable interrupt or exception by setting a breakpoint to gain control. Their defense method is to make use of the stack to store a part of the checksum. They claimed that all CPUs automatically save some state on the stack when an interrupt or exception occurs. If the stack pointer is pointing to the checksum that is on the stack, any interrupt or exception will cause the processor to overwrite the checksum. However, this stack trick does not work for DRSA because the status information will be saved in debug registers instead of on the stack if the exception is generated by a hardware breakpoint.

To confirm this conclusion, we obtained the address of the stack pointer before and after the execution of the exception handlers to verify whether it has changed. We conducted the experiment on the ptrace-based DRSA. The experiment platform used in our experiment is a PC with a 3.41 GHz Intel i7-6700 processor running the Ubuntu 18.04 operating system. By using *PTRACE_GETREGS* request, the DRSA process (the "tracer") could copy the general-purpose or floating-point registers of the attestation process (the "tracee") to its address data. By accessing the RSP register, we can obtain the stack pointer. In order to compare the difference between hardware breakpoints and software breakpoints, we implemented a process of gaining control from the attestation function through a software breakpoint. As shown in Fig. 6, the first instruction of the attestation function is replaced with an INT3 instruction. This instruction is a one-byte instruction defined to temporarily replace an instruction in a running program in order to set a software breakpoint. When the attestation code tries to run, a debug exception will be raised, thereby gaining control. We also record the stack pointer before and after the software breakpoint is triggered. The experimental results are shown in Table 3.

The experimental results illustrate that exceptions raised by the data read/write hardware breakpoint and single stepping in DRSA do not cause the stack pointer to move. In a comparative experiment, however, the stack pointer

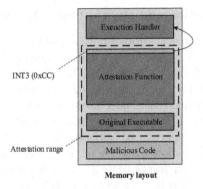

Memory layout

Fig. 6. Gaining control from the attestation function through a software breakpoint

Table 3. The value of the stack pointer. Stack pointer before the breakpoint triggers (sp/ Before.bp), stack pointer after the breakpoint triggers (sp/ After.bp) and stack pointer after the single stepping (sp/ After.ss).

Breakpoint Type	sp/ Before.bp	sp/ After.bp	sp/ After.ss
Software Breakpoint	$0 \times 7FFF6F6DAB78$	$0 \times 7FFF6F6DAB60$	–
Hardware Breakpoint (DRSA)	$0 \times 7FFCDE29DDB8$	$0 \times 7FFCDE29DDB8$	$0 \times 7FFCDE29DDB8$

points to different locations before and after the execution of INT3 instruction. This will cause the checksum to be overwritten and further cause the manipulations to be detected by the attestation function. We conclude that DRSA can escape from the detection of the existing attestation routine on the x86 platform.

7   Related Work

A. Software Attestation

SWATT [17] is an early software attestation scheme based on challenge-response protocol. It requires a strict running time of the attestation procedure to prevent an attacker from redirecting the memory regions. The fact that SWATT relies on optimized function execution and a strict time limit is considered a drawback. Pioneer [16] is a software attestation scheme similar to SWATT, focusing on the x86 platform. Considering that the target devices are equipped with a large amount of memory, Pioneer is self-checksumming and establishes a dynamic root of trust to invoke a trusted executable. Pioneer also requires strict time constraints to detect memory copy attacks. Distributed [21] uses memory filling to overcome the time constraints imposed by SWATT. The free memory of the device is filled with randomness before deployment, preventing the attacker from having empty memory to evade detection during attestation. However, schemes based on memory-filling techniques are threatened by compression attacks, which compress the original program of the victim device, opening up space for malicious code.

B. Hardware Attestation and Hybrid-based Attestation

To address limited security protections of software attestation schemes, hardware-based approaches rely on trusted computing architectures such as TPM [11], ARM TrustZone [4], Sancus [14]. Despite their strong security guarantees, the requirement for costly customized hardware that cannot be accommodated in small IoT platforms makes hardware-based protocols incompatible with many low-end devices. To this end, hybrid-based solutions, such as TyTAN [5], leverage the best properties of software-based and hardware-based RA approaches to establish Root-of-Trust by relying on minimal hardware assumptions. In particular, hybrid solutions require modification of device hardware to ensure atomic and secure code execution of RA protocols. One recent work that aims to fill the gap between software-based and hybrid-based RA schemes is SIMPLE [3], a hypervisor-based RA scheme for resource-constrained IoT devices. SIMPLE relies on a software-based memory isolation technique called Security MicroVisor (SμV) [2]. SμV shields a software-based Trusted Computing Module (TCM) from untrusted application software using selective software virtualization and assembly-level code verification. However, due to the runtime safety checks, this approach introduces increased execution time. Moreover, SμV is considered memory-safe and crash-free, but it has not been fully verified yet. HAtt [1] is a hybrid remote attestation, which ensures the high availability of IoT devices during the software attestation process. The proposed attestation technique uses physical unclonable functions (PUFs) to protect the secrets of an IoT device from physical attacks. The security analysis shows that the proposed attestation technique can effectively detect roving malware. RATA [8] is a provably secure approach to detect transient malware that infects a device (by modifying its binary), performs its nefarious tasks, and erases itself before the next attestation. SARA [9] is an attestation protocol that aims to attest a large number of devices. SARA exploits asynchronous communication capabilities among IoT devices in order to attest to a distributed IoT service they execute. LIRA-V [19] is a lightweight system for performing remote attestation between constrained devices using the RISC-V architecture. It uses read-only memory and the RISC-V Physical Memory Protection (PMP) primitive to build a trust anchor for remote attestation and secure channel creation. SCRAPS [15] is a collective RA scheme that achieves scalability by outsourcing verifier duties to a smart contract and mitigates DoS attacks against both provers and verifiers.

Nevertheless, most of the works focus on the integrity of the lightweight embedded devices, and the proposed attestation protocols need to check the whole memory of the target devices. However, with the development of computer hardware, even some IoT devices are equipped with a large amount of memory and installed operating system. The memory of these devices is also at risk of malicious tampering. DRSA focuses on challenging the remote attestation schemes on devices with more processing power and more memory. By utilizing debug registers, DRSA can relocate itself and escape detection from existing software attestation schemes.

8   Conclusion and Future Work

In this paper, we presented DRSA, a new specific attack against existing software attestation techniques on the x86 platform. DRSA circumvents attestation by relocating itself and restoring the contents before it is measured. It leverages debug registers provided by the debugging architecture to monitor the beginning and the end of an RA measurement, erase itself, and restore the original program memory contents before attestation. We designed and implemented the prototype of DRSA under the Windows and Linux operating systems, respectively. From our experience, we can conclude that DRSA incurs very little attack overhead, which is difficult to detect by the existing time-based attestation schemes on the x86 platform. We also stress that an attacker can hide malicious code using debug registers, which is not taken into account by existing attestation schemes. We argue that attestation schemes should pay more attention to high-performance devices with debug architectures and more memory. For future work, we will investigate software-based attestation protocols that can guarantee security for devices with operating systems and more complex mechanisms.

References

1. Aman, M.N., et al.: HAtt: hybrid remote attestation for the internet of things with high availability. IEEE Internet Things J. **7**(8), 7220–7233 (2020)
2. Ammar, M., Crispo, B., Jacobs, B., Hughes, D., Daniels, W.: Sμv-the security microvisor: a formally-verified software-based security architecture for the internet of things. IEEE Trans. Dependable Secure Comput. **16**(5), 885–901 (2019)
3. Ammar, M., Crispo, B., Tsudik, G.: Simple: a remote attestation approach for resource-constrained IoT devices. In: 2020 ACM/IEEE 11th International Conference on Cyber-Physical Systems (ICCPS), pp. 247–258. IEEE (2020)
4. ARM, A.: Security technology building a secure system using trustzone technology (white paper). ARM Limited (2009)
5. Brasser, F., El Mahjoub, B., Sadeghi, A.R., Wachsmann, C., Koeberl, P.: TyTAN: tiny trust anchor for tiny devices. In: Proceedings of the 52nd Annual Design Automation Conference, pp. 1–6 (2015)
6. Carpent, X., Rattanavipanon, N., Tsudik, G.: Remote attestation of IoT devices via smarm: shuffled measurements against roving malware. In: 2018 IEEE International Symposium on Hardware Oriented Security and Trust (HOST), pp. 9–16. IEEE (2018)
7. Castelluccia, C., Francillon, A., Perito, D., Soriente, C.: On the difficulty of software-based attestation of embedded devices. In: Proceedings of the 16th ACM Conference on Computer and Communications Security, pp. 400–409 (2009)
8. De Oliveira Nunes, I., Jakkamsetti, S., Rattanavipanon, N., Tsudik, G.: On the TOCTOU problem in remote attestation. In: Proceedings of the 2021 ACM SIGSAC Conference on Computer and Communications Security, pp. 2921–2936 (2021)
9. Dushku, E., Rabbani, M.M., Conti, M., Mancini, L.V., Ranise, S.: SARA: secure asynchronous remote attestation for IoT systems. IEEE Trans. Inf. Forensics Secur. **15**, 3123–3136 (2020)

10. Eldefrawy, K., Rattanavipanon, N., Tsudik, G.: Hydra: hybrid design for remote attestation (using a formally verified microkernel). In: Proceedings of the 10th ACM Conference on Security and Privacy in wireless and Mobile Networks, pp. 99–110 (2017)
11. Group, T.C.: Trusted platform module (TPM) (2017). http://www.trustedcom putinggroup.org
12. Guide, P.: Intel® 64 and ia-32 architectures software developer's manual. Volume 3B: System programming Guide, Part **2**(11), 0–40 (2011)
13. Hao, S., et al.: Deep reinforce learning for joint optimization of condition-based maintenance and spare ordering. Inf. Sci. **634**, 85–100 (2023)
14. Noorman, J., et al.: Sancus: low-cost trustworthy extensible networked devices with a zero-software trusted computing base. In: 22nd USENIX Security Symposium (USENIX Security 13), pp. 479–498 (2013)
15. Petzi, L., Yahya, A.E.B., Dmitrienko, A., Tsudik, G., Prantl, T., Kounev, S.: {SCRAPS}: scalable collective remote attestation for {Pub-Sub}{IoT} networks with untrusted proxy verifier. In: 31st USENIX Security Symposium (USENIX Security 22), pp. 3485–3501 (2022)
16. Seshadri, A., Luk, M., Shi, E., Perrig, A., Van Doorn, L., Khosla, P.: Pioneer: verifying code integrity and enforcing untampered code execution on legacy systems. In: Proceedings of the Twentieth ACM Symposium on Operating Systems Principles, pp. 1–16 (2005)
17. Seshadri, A., Perrig, A., Van Doorn, L., Khosla, P.: SWATT: software-based attestation for embedded devices. In: IEEE Symposium on Security and Privacy, 2004. Proceedings. 2004, pp. 272–282. IEEE (2004)
18. Shaneck, M., Mahadevan, K., Kher, V., Kim, Y.: Remote software-based attestation for wireless sensors. In: Molva, R., Tsudik, G., Westhoff, D. (eds.) ESAS 2005. LNCS, vol. 3813, pp. 27–41. Springer, Heidelberg (2005). https://doi.org/10.1007/11601494_3
19. Shepherd, C., Markantonakis, K., Jaloyan, G.A.: LIRA-V: lightweight remote attestation for constrained RISC-V devices. In: 2021 IEEE Security and Privacy Workshops (SPW), pp. 221–227. IEEE (2021)
20. Yang, X., et al.: Towards a low-cost remote memory attestation for the smart grid. Sensors **15**(8), 20799–20824 (2015)
21. Yang, Y., Wang, X., Zhu, S., Cao, G.: Distributed software-based attestation for node compromise detection in sensor networks. In: 2007 26th IEEE International Symposium on Reliable Distributed Systems (SRDS 2007), pp. 219–230. IEEE (2007)
22. Zhang, N., Tan, Y.A., Yang, C., Li, Y.: Deep learning feature exploration for android malware detection. Appl. Soft Comput. **102**, 107069 (2021)
23. Zhang, Q., et al.: A hierarchical group key agreement protocol using orientable attributes for cloud computing. Inf. Sci. **480**, 55–69 (2019)
24. Zhu, H., Tan, Y.A., Zhu, L., Zhang, Q., Li, Y.: An efficient identity-based proxy blind signature for semioffline services. Wireless Commun. Mobile Comput. **2018**(2), 1–9 (2018)

Fuzz Testing of UAV Configurations Based on Evolutionary Algorithm

Yuexuan Ma[1], Xiao Yu[1(⊠)], Yuanzhang Li[2], Li Zhang[3], Yifei Yan[1], and Yu-an Tan[2]

[1] School of Computer Science and Technology, Shandong University of Technology,
Zibo 255049, Shandong, China
yuxiao8907118@163.com
[2] School of Computer Science and Technology, Beijing Institute of Technology,
Beijing 100081, China
[3] Department of Media Engineering, Communication University of Zhejiang,
Hangzhou 310018, Zhejiang, China

Abstract. With the widespread application of Unmanned Aerial Vehicle (UAV) technology, its security issues have also attracted much attention, among which the configuration attack against the UAV flight control system is one of the current research hotspots. Attackers always upload seemingly normal configuration combinations and cause an imbalance in the UAV state by exploiting configuration item verification vulnerabilities. This paper accumulates flight data through simulation, generates configuration combinations within the security range using differential evolution-based fuzz testing, uses neural networks to guide configuration item variants, and applies these configuration combinations to the AutoTest of UAV flight control systems. The experimental results show that the configuration combinations generated by fuzz testing can guide the UAV to course deviation, spin, crash and other unstable states; the code coverage and function coverage of the position and attitude code library base in the flight control system have also reached a high level.

Keywords: UAV · Configuration Security · Fuzz Testing · Differential Evolution · Neural Network · Code Coverage

1 Introduction

UAV is a type of aircraft that does not require direct maneuvering by personnel. Due to its excellent reconnaissance and strike capabilities, it was initially used for military purposes at first. In recent years, technological advances and cost reductions have led to the widespread use of UAVs in civil applications such as agricultural engineering, environmental monitoring, commercial performances, and aerial photography. However, as the functionality of UAV flight control systems continues to expand and the complexity of the structure continues to increase, configuration errors can accumulate throughout the system. These problems may lead to the unavailability of system functions or severe safety incidents.

© ICST Institute for Computer Sciences, Social Informatics and Telecommunications Engineering 2024
Published by Springer Nature Switzerland AG 2024. All Rights Reserved
J. Chen and Z. Xia (Eds.), BlockTEA 2023, LNICST 577, pp. 41–56, 2024.
https://doi.org/10.1007/978-3-031-60037-1_3

Fuzz testing is a testing technique to deal with such software security issues. By injecting a large number of unexpected test cases into the target program, fuzz testing seeks to cover all the running states of the program to discover as many configuration errors in the system as possible. It has the advantages of relatively low cost, high efficiency in finding errors, and automatic execution. Common fuzz testing tools include AFL [1], Peach, Radamsa, etc. These tools generate test cases by changing existing data samples or modeling program inputs, and adjust test cases according to program status feedback to detect program errors with loopholes. Usually, flight control system has thousands of configuration items, and the configuration combination will form a huge input space; and there is currently no unified judgment condition to determine whether the flight control system is faulty, so the above fuzz testing tools are mainly used for traditional software such as applications, and cannot be directly applied to embedded dedicated systems such as ArduPilot [2].

In response to this problem, Kim et al. used the control-guided testing method to discover the input verification vulnerability of the UAV flight control system [2]; and Han et al. proposed a fuzz testing tool LGDFuzz for the UAV flight control system [3], the test cases are generated by the genetic algorithm, and the effect of the fuzz testing is judged by the state of the drone. However, the above-mentioned fuzz testing methods for UAVs do not take the coverage rate as the guidance of the fuzz testing. Using the coverage index instead of observing the state of the UAV can evaluate the results of the fuzz testing more intuitively and accurately. Therefore, this paper proposes a fuzz testing method based on differential evolutionary algorithm to fuzz test some configuration items of ArduCopter, a flight control system under ArduPilot, and analyzes the configuration errors and security loopholes in the system using the coverage rate and the failure rate as the evaluation indexes. The main contributions of this paper are as follows:

(1) For mainstream fuzz testing tools only designed for the scope of application of general-purpose software, a fuzz testing tool adapted to the ArduCopter flight control system is proposed. Evolutionary algorithm fuzz testing for finding configuration errors in flight control systems.
(2) For the problem that the combination of fuzz testing based on the differential evolutionary algorithm and UAV flight control system leads to high computational cost, a neural network model TALA is established that integrates convolutional neural network, recurrent neural network and attention mechanism. This model predicts the state of UAV to accelerate the computation of fitness function in differential evolutionary algorithm and improve the efficiency of fuzz testing.

2 Related Works

2.1 Attack Methods Against UAVs

The widespread use of UAVs exposes them to many attacks, including physical attacks on sensors, network attacks on communication protocols, and software attacks on flight control systems. At the physical attack level, laser blinding attacks against vision sensors [4, 5] may lead to incorrect decisions by the UAV, which in turn may lead to more serious consequences; acoustic resonance attacks against UAV navigation sensors [6] can cause the UAV to lose position data and thus crash uncontrollably; spoofing attacks against GPS [7, 8] can force UAVs to receive false navigation information, thereby inducing dangerous behavior. At the level of communication protocol security, as the MAVLink communication protocol commonly used by UAVs does not support encrypted communication and authentication authorization, the attackers can intercept and modify MAVLink information between the drone and the ground station, and easily launch man-in-the-middle attacks or replay attacks. Kwon et al. [9] performed ICMP flooding attack and malicious packet injection attack on UAVs by analyzing the MAVLink protocol vulnerability, causing the mission UAVs to stop and hover due to deleting mission information. At the software security level, the lack of validation of configuration item inputs by the flight control system makes it possible for seemingly normal inputs to cause anomalies in the UAV, i.e., an input validation vulnerability exists. RVFuzzer [2] proposed by Kim et al. looks for input validation errors in the flight control system through control-guided input variants, and LGDFuzz [3] proposed by Han et al. similarly exploits input validation vulnerabilities to perform configuration attacks on UAVs. The work in this paper is also developed based on this classical vulnerability to attack the UAV flight control system by mutating the configuration items of the UAV flight control system within the legitimate range.

2.2 Fuzz Testing for UAVs

As an embedded special system, fuzz testing for UAVs has difficulties such as complex data interaction, high environmental dependency, and strict safety and reliability requirements. As a result, general-purpose fuzz testing tools that cannot be directly used for UAV flight control system. However, their seed mutation methods, seed selection and other ideas can still be used as a reference. As one of the most representative fuzz testing tools, AFL [1] and its improved versions are mutation-based file fuzz testing tools, which generate test cases through genetic algorithms and track the execution path of the code under test by using instrumentation technology. Record the code coverage of the input samples, and use them as feedback to mutate the input samples to improve the coverage, thereby enhancing the possibility of finding vulnerabilities. Honggfuzz [10] is also a mutation-based fuzz testing tool that generates test cases and detects vulnerabilities in target programs through dynamic binary instrumentation and automated symbol execution techniques. Peach is a model-based file fuzz testing tool that automatically generates random test cases that meet the target program's expectations according to the input data's format and structure and records the coverage results for analysis. This method can ensure test cases' passability and improve fuzz testing's efficiency and accuracy. LibFuzzer [11] is a memory-based fuzz testing tool that uses instrumentation technology to track code coverage. By establishing each fuzz testing in the same process, the injection of the test case does not need to wait for the restart of the target program, which greatly improves the efficiency of the fuzz testing.

The above general fuzz testing tools' design ideas and implementation schemes have laid the foundation for applying fuzz testing on UAVs. RVFuzzer [8] uses control-guided input mutators to generate test cases and uses the detection results of the UAV's state as feedback to fuzz testing the flight control system to detect input validation vulnerabilities. LGDFuzz [9] uses genetic algorithm-based fuzz testing to automatically detect errors in UAV configuration items for range specification errors and uses multi-objective optimization to calculate the most suitable parameter value range for each configuration item. PGFuzz [12] proposes a policy-oriented fuzz testing framework for robotic vehicles, which expresses temporal logic formulas with time constraints, and uses this constraint as a guiding metric for fuzzing. In this paper, ArduCopter, a multi-rotor UAV flight control system under ArduPilot, is fuzzed for its configuration items using the differential evolutionary algorithm, and the coverage rate and failure rate are used as evaluation metrics for fuzz testing.

2.3 Neural Networks for Fuzz Testing

Traditional fuzz testing usually uses randomly generated test cases to test software programs. This method can cover a relatively complete test space, but it is difficult to guarantee the efficiency and accuracy of the test. With the rise of machine learning, fuzz testing technology integrated with deep learning has gained a wide range of development space. NEUZZ [13] uses a feed-forward neural network to learn a smooth approximation of the branching behavior of the target program to generate effective test cases; and uses a gradient-guided search strategy to guide mutation positions to maximize the detection of errors in the target program. Faster Fuzzing [14] uses Generative Adversarial Networks (GAN) as AFL's seed reinitialization strategy, which can find more unique code paths than LSTM or random enhancement strategies, and enhances the fuzzing effect of AFL. SmartSeed [15] proposed by Lyu et al. uses Wasserstein Generative Adversarial Networks (WGAN) to generate valuable seed files, which are used as the input of AFL for fuzz testing. LGDFuzz [9] uses LSTM to calculate the fitness function in the fuzz testing based on the genetic algorithm to estimate the state of the UAV. In this paper, a neural network model combining TCN, LSTM and Attention is established in the evolutionary algorithm, which can predict the state of the drone with a lower error and accelerate the fitness function calculation.

3 UAV Fuzz Testing Framework

3.1 UAV Fuzz Testing Overall Framework

This paper proposes a fuzz testing method based on coverage and failure rate evaluation. First, configure the UAV mission file and select the configuration items to participate in the fuzz testing. Then, the flight log of the UAV is collected by the emulator in the way of simulated flight, and the characteristic data of the log are extracted to construct a data set for the neural network and fuzz testing. The differential evolutionary algorithm-based fuzz testing uses a neural network to calculate fitness function and generates many deformed configuration items using a mutation data set. Finally, the results of fuzz testing are applied to simulation verification, and the coverage and failure rates of specific components are calculated to evaluate the performance of fuzz testing and guide the selection of configuration items. The overall framework of the UAV fuzz testing method is shown in Fig. 1.

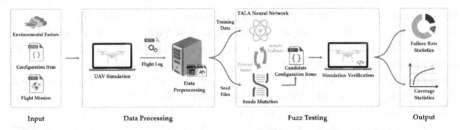

Fig. 1. UAV fuzz testing overall framework.

3.2 UAV Simulation

Operating Platform and Flight Mission. Both the acquisition of UAV flight logs and the verification of configuration items need to be carried out on the flight control system. In this paper, SITL simulator with lower cost but higher efficiency and safety is selected to simulate and run UAVs. Different configuration items are randomly created within the ArduPilot published configuration item safety range [16], and the AVC2013 [17], FitCollection, Copter-Mission, and SDUT-Map flight missions are repeatedly executed with these configuration item groups to collect flight logs. Among them, AVC2013 is the flight route used in SparkFun's fifth Autonomous Vehicle Competition. The UAV needs to take off independently, pass through the designated waypoint and land in the landing area to earn points; FitCollection is the flight route used by LGDFuzz [9], which is an improved version of AVC2013; Copter-Mission is the flight route used by the ArduPilot autotest framework, taking off from the airport of the Canberra Model Aircraft Club and returning to the airport after a simple flight. All of the above are public flight routes, while SDUT-Map belongs to this paper's own planned flight routes, aiming to evaluate UAV fuzz testing's adaptability to complex routes. During log collection process, abnormal logs are eliminated through data consistency inspection to prevent the training data set of the neural network from being polluted.

Configuration Item Selection. ArduPilot achieves specific functions by decoupling different system libraries, including core library, navigation library, control library, sensor library and motor library, etc. Different configuration items control various libraries. Due to the large number of configuration items involved in the system library and the lack of correlation, this article only selects the configuration items that affect the flight status of UAVs under the automatic flight state, while the configuration items of non-closely related components such as remote control and camera are not considered in the fuzz testing. ArduPilot mainly uses the position controller and attitude controller to control the flight state of the UAV, and the related configuration items are located primarily in the AC_AttitudeControl library. In the position controller, the horizontal position controller converts the position error into a target horizontal velocity, and the velocity controller converts the velocity error into a desired acceleration, which is then converted into a desired tilt angle and sent to the attitude controller. In the position controller, the horizontal position controller converts the position error into a target horizontal velocity, and the velocity controller converts the velocity error into a desired acceleration, which is then converted into a desired tilt angle and sent to the attitude controller. The vertical position controller converts the position error into a target vertical velocity, the velocity controller converts the velocity error into a desired acceleration, and the acceleration PID controller converts the acceleration error into a desired throttle, which is then sent to the attitude controller. The position controller is controlled by three configuration items of speed loop, position loop and acceleration loop. In this paper, 14 configuration items involving PID gain and feedforward gain in each loop are selected. In the attitude controller, the AP controller converts the error between the target angle and the actual angle into a desired rotation rate. Then the attitude PID controller converts the rotation rate error into high-level motor commands to control the state of the UAV. The attitude controller is controlled by the configuration items of the angular rate loop and the angle loop. In this paper, 23 configuration items involving PID gain, feedforward gain and limiter are selected in each loop, which have the most direct impact on the position and attitude of the UAV in AUTO flight mode.

3.3 Data Preprocessing

The UAV log data needs to be preprocessed to improve the neural network's performance and the effect of fuzz testing. After collecting enough flight logs, extract three angular states (Roll Angle, Pitch Angle, Yaw Angle), three angular rates (Roll Angle Rate, Pitch Angle Rate, Yaw Angle Rate), 12 sensor data (Three-axis Accelerometer Original Value, Three-axis Gyroscope Original Rotation Rate, Three-axis Compass Original Magnetic Field Value, Three-axis Main Accelerometer Output Standard Deviation) and 42 configuration items, merged into a log dataset in CSV format, and create a copy of the dataset. A normalization operation is applied to the features of one of the datasets so that different features have the same scale to optimize the speed of gradient descent to find the optimal solution. Finally, the data set is transformed into a supervised learning problem. The training set and test set are divided for subsequent neural network training. After sampling another data set, it is directly converted into a supervised learning data set, which is used to build a seed pool for fuzz testing.

3.4 Fuzz Testing

General Framework of Fuzz Testing. Fuzz testing is a method of discovering vulnerabilities by providing many malformed inputs to the target system and observing the abnormal results exhibited by the system. Relatively low cost, high efficiency of configuration error mining, and automatic execution make fuzz testing a common method for software black-box, white-box or gray-box testing.

According to the acquisition method of test samples, fuzz testing can be divided into the mutation mode based on changing the existing data samples and the generation method based on program input modeling. However, the traditional method of randomly generating test samples has a poor purpose, leading to low program testing efficiency. In the fuzzification of configuration items, this paper uses the Different Evolution Algorithm (DE) [18, 19] to mutate the configuration items, which is a heuristic optimization algorithm based on the genetic algorithm, and its process is shown in Algorithm 1. Firstly, the default value of each configuration item is used as the initial population, and each individual in it corresponds to a group of configuration items. Then calculate the fitness of individuals in the population, perform evolutionary operations of selection, mutation, and crossover on individuals, and continue to iterate until the termination condition. The last generation population is used as the target configuration item population to construct the output of the fuzz testing.

Algorithm 1: Differential evolution algorithm.

Input: $P(G)$, MAX

Output: $P(G_{MAX})$

1: **while** $G < MAX$ **do**

2:   $f(P(G)) \leftarrow Fitness(P(G))$

3:   $P(G) \leftarrow$ Selection($f(P(G))$, $P(G)$)

4:   $P(G) \leftarrow$ Mutations($P(G)$)

5:   $P(G) \leftarrow$ Crossover($P(G)$)

6:   $f(P(G)) \leftarrow Fitness(P(G))$

7:   $P(G + 1) \leftarrow$ Environmental Selection($f(P(G))$, $P(G)$)

8: **end while**

Population Initialization. The differential evolution algorithm constructs an initial candidate solution population $P(0)$, which contains a number of individuals $X_{i,j}$ composed of chromosomes, each individual obeys the random deviation of the standard normal distribution plus the uniform distribution between the lower bounds, and each chromosome in the individual All are candidate solutions for the current problem. The populations $P(0)$ and individuals $X_{i,j}$ are shown below:

$$\begin{cases} \left\{ P(0) | x_{i,j}^{(L)} \leq x_{i,j}(0) \leq x_{i,j}^{(U)}; i = 1, 2, \cdots, MAX; j = 1, 2, \cdots, D \right\} \\ x_{i,j}(0) = RND[0, 1] * \left(x_{i,j}^{(U)} - x_{i,j}^{(L)} \right) + x_{i,j}^{(L)} \end{cases} \quad (1)$$

where i is the individual number, j represents the j th dimension; $x_{i,j}^{(L)}$ and $x_{i,j}^{(U)}$ represent the lower and upper bounds of the i th individual in the j dimension; MAX represents the population size; D represents the independent variable dimension; $RND[0, 1]$ represents a random number in the interval $[0, 1]$. The defined chromosome's objective function value in the candidate solution population is the configuration item's default value.

Fitness Function. The fitness function is an important evaluation index for the survival of the fittest, which directly affects the convergence speed of the evolutionary algorithm and whether it can find the optimal solution. The fitness function can be expressed as *Fitness* = |*PredictiveValue* − *TrueValue*|, which is the absolute value of the deviation between the predicted and true values of the three angular states (Roll, Pitch, Yaw) and three angular rates (RateRoll, RatePitch, RateYaw) of the UAV. In this paper, the neural network is used to calculate the predicted value of the UAV state, and the specific information is shown in Sect. 3.5. The larger the fitness function is, the more likely this group of configurations causes the possible state of the UAV to deviate from the real state, and the more likely it is an abnormal configuration item group, so it is more likely to be saved as the optimal individual in the next generation population.

Evolutionary Operations. Evolutionary operations include selection, mutation, crossover, and environmental selection.

Selection. The selection operator adopts the elite selection method, directly selecting a certain number of individuals with high fitness and directly participating in the subsequent evolutionary operation without modification. In contrast, the unselected individuals are directly discarded and will not enter the evolutionary iteration.

Mutations. The difference vector difference is scaled after different individuals are obtained by selection. Then, the vector is fused with the individual to be mutated to obtain a new mutated individual. Aggressive mutation probabilities can lead to many invalid inputs, while more conservative mutation probabilities can avoid falling into local optimal solutions. The operation of differential mutation is as follows:

$$V_{i,g} = x_{i,g} + F * \left[\left(x_{r1,g} - x_{r2,g} \right) + \left(x_{best,g} - x_{i,g} \right) \right] \tag{2}$$

where $V_{i,g}$ is the mutant individual, g represents the evolution algebra, and F is the variation scaling factor, which is used to control the scaling of the difference vector; $r1$ and $r2$ are unequal random numbers in the interval $[0, NP]$, $x_{r1,g}$ is a randomly selected individual in the population, $x_{r2,g}$ is the randomly selected individual in the current population and the external archive collection, and $x_{r1,g} - x_{r2,g}$ constitutes the difference vector; $x_{best,g}$ is the best individual in the current population, $x_{i,g}$ is the parent individual. After the differential mutation operation, each parent individual can generate a mutant individual.

Crossover. To further increase the diversity of individuals, the binomial distribution crossover method crosses the mutant individuals with the parent individuals to generate new test individuals. The operation of the binomial crossover is as follows:

$$U_{i,j}(g) = \begin{cases} V_{i,j}(g); & if \ rand(0,1) \leq XVOR \\ x_{i,j}(g); & if \ rand(0,1) > XVOR \end{cases} \tag{3}$$

where $U_{i,j}(g)$ is a new test individual, and $V_{i,j}(g)$ is the j th parameter value of the i th configuration in the next-generation individual after crossover, $x_{ij}(g)$ is the j th parameter value of the i th configuration in the g th generation individual before crossover, and $XVOR$ is the crossover probability.

Environmental Selection. One-to-One Survivor Selection (OTOS) is performed between the test individuals and the parent individuals in index order, and the next generation retains the elite individuals with higher fitness to generate a new population generation. The operation of the environmental selection is shown below:

$$X_i(g+1) = \begin{cases} U_i(g); & if \ Fitness(U_i(g)) \leq Fitness(X_i(g)) \\ X_i(g); & if \ Fitness(U_i(g)) > Fitness(X_i(g)) \end{cases} \tag{4}$$

where $X_i(g+1)$ is the selected next-generation individual, $U_i(g)$ is the test individual, $X_i(g)$ is the parent individual, and *Fitness* is the fitness function.

Generating New Generation Populations. If the maximum evolutionary generation or the convergence condition of the algorithm is not reached, the generated new generation population will continue to the next round of evolution; otherwise, some individuals with the highest fitness will be selected from the last generation population to construct the configuration population as the test case output by the fuzz testing.

3.5 UAV Status Prediction

The ArduPilot flight control system collects sensor data such as Inertial Measurement Unit (IMU), GPS, and Compass and uses the Extended Kalman Filter (EKF) to estimate information such as the position, attitude, and velocity of the UAV. In the fitness function of the evolutionary algorithm, to efficiently calculate the predicted value of each state of the UAV, we use a TCN-Attention-LSTM-Attention (TALA) neural network to estimate the angular state and angular rate of the UAV. The input layer of the model consists of 3 angular states of the UAV, three angular rates, 12 sensor data, and 42 configuration items for the first four-time steps. In the hidden layer, the TCN layer extracts the UAV state and configuration item features in the time dimension and updates the weights of different feature information by the first layer of Attention. The LSTM prediction layer further extracts time series features and then updates the different features by the second layer of Attention. After paying attention to the information, the output layer outputs the maximum conditional probability prediction of the UAV's three angular states and three angular rates in the next time step. The overall framework of the neural network is shown in Fig. 2.

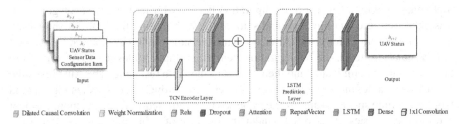

Fig. 2. The general framework of neural networks.

The Temporal Convolution Network (TCN) [20] uses Casual Dilated Convolutions to model the time series of UAV angular states, angular rates, sensor data, and configuration item information. In which causal convolution only performs one-way processing of past and current moment data and does not introduce future information, avoiding the traditional CNN data leakage problem; the dilated convolution can flexibly adjust the receptive field of the neural network, so that the network can handle the long-term dependence characteristics of the flight log sequence. The weights of the feature information affecting the UAV state change dynamically in different time periods, and the Attention mechanism [21] enables the neural network to assign higher weights to the focused information related to the output UAV state in the input UAV state, sensor information and configurations, thus increasing the network's attention to the focused information and reducing the computational burden of the network. LSTM uses input gate, forget gate and output gate mechanisms to deal with long-term dependencies in sequences by selectively forgetting and updating past information.

The neural network model that mixes TCN, LSTM and Attention can more comprehensively capture the long-term and short-term dependencies in the time series, and focus on important features related to the UAV state in the sequence, thereby improving the accuracy and stability of time series modeling.

3.6 Simulation Verification

Simulation Validation. Although each configuration item constructed by the fuzz testing is within the range specified by ArduPilot, it may be very deformed data after combination, which will cause various abnormal states of the UAV. We use simulation to verify whether these configuration items will cause different abnormal UAV states. The possible exception states of the UAV and their meanings are described in Sect. 3.6. As a flight control system, ArduPilot has error reporting and protection programs for abnormalities in the UAV. Unexpected configuration items can explore more edge codes and achieve higher code coverage.

The simulation verification module first uploads the configuration item groups constructed by the fuzz testing to the SITL simulation environment, uses the sim_vehicle script to start ArduCopter and cyclically verifies each group of configuration items, and checks the abnormal status of the UAV by listening to the MAVLink message sent by the UAV. This test method needs to restart the SITL simulator to verify the next configuration items, which is less efficient for testing. This script only repeats the same flight mission with a new set of configuration items, covering fewer test categories, so it is only used to count the abnormal state of the UAV during the flight. It should not be used for coverage detection. After verifying the state of the UAV caused by each set of configuration items, the configuration item group that enables the UAV to perform tasks normally is eliminated and then passed to the AutoTest suite. The AutoTest suite creates repeatable unit tests and functional tests for ArduCopter. This testing method reduces the time spent on running the same scenario repeatedly. Testing is more efficient and more categories are covered. Under the premise of using the same configuration items, we combine the two configuration item verification methods of sim_vehicle and AutoTest to comprehensively calculate the code coverage, function coverage and failure rate of the ArduPilot related code base.

Abnormal State of the UAV. Unexpected configuration items may lead to abnormal states of the UAV, and the quality of test cases generated by fuzz testing can be measured by monitoring these abnormal states. High-quality test cases can guide UAVs to appear in various abnormal states.

The abnormal state of the UAV can be analyzed by downloading log files or monitoring the communication data between the UAV and the ground station. However, the log files of the UAV usually need to be downloaded after a complete flight, which has great limitations in time. By monitoring the MavLink communication protocol and analyzing the heartbeat packets transmitted by the UAV to the ground station in real time, the abnormal state of some UAVs can be judged immediately. The state categories of UAVs are shown in Table 1.

Table 1. Status categories of UAVs.

UAV Status	Implications
Pass	If the UAV can complete the scheduled flight mission with a relatively normal attitude, this flight mission can be defined as Pass
Timeout	Incorrect configuration items may cause the UAV to hover or stay in a certain position. When not idling on the ground or hovering in the air, if the moving distance within a certain period is less than the threshold, it will be considered as Timeout
PreArm Failed	Flight control system will refuse to unlock the motors when it detects sensor abnormalities in sensors or some configurations before takeoff
Potential Thrust Loss	When certain configurations are wrong, the load is too large, or the ambient wind speed is too high, even if the throttle is saturated, the UAV cannot reach the desired state
Course Deviation	Incorrect configuration items may cause the UAV in Auto mode to deviate from the established route
Yaw Imbalance	The higher the yaw imbalance, the weaker the UAV's control over the yaw action, and the more likely it is to fall into a state of swinging left and right in the yaw direction. Conversely, UAVs have more room for stability control
Spin	The UAV will fall into a fast spin when the yaw imbalance exceeds a certain threshold. This abnormal state usually cannot be controlled or stopped
Crash	Abnormal conditions such as excessive landing speeds, consistently exceeding angular limits, loss of thrust, yaw imbalance, and spin can all lead to a collision or crash of the UAV

3.7 Result Statistics

Failure discovery rate refers to the ratio of the number of test cases that can trigger vulnerabilities and errors to the total number of test cases. It is one of the important indicators to measure the effect of fuzz testing. We counted the number and proportion of the test cases generated by the fuzz testing that can cause the abnormal state of the UAV described in Sect. 3.6 to judge the quality of the test cases.

Code coverage and function coverage are common metrics for fuzz testing methods. The higher the coverage, the more comprehensive the testing of the program, the more likely to find defects in the code. We use GCOV as a statistical tool for code coverage. When compiling the source code of the target program, we insert probes into the entry and exit of the basic block to determine whether the basic block is covered. After running enough tests, generate a source file of coverage statistics in GCOV format and then use LCOV to convert it into a visual coverage report in HTML format. The coverage report reflects the code coverage and function coverage of specific files. It even includes the execution times of each line of code, which can intuitively evaluate the effect of fuzz testing.

4   Experiments

4.1   Experiment Environment

The hardware environment of this experiment is Intel i7-12700KF CPU, NVIDIA RTX 3080 GPU, 64 GB RAM; the software environment is Ubuntu 22.04 operating system, ArduPilot V4.3.7 open source autopilot software, Python 3.10.6 development language, TensorFlow 2.12.0 deep learning framework.

4.2   Experimental Evaluation Index

Neural Network Evaluation Index. The deviation between true and predicted values usually evaluates the regression problem. This paper uses Mean Square Error (MSE) as the loss function of the neural network. The formula for MSE is shown below:

$$MSE = \frac{1}{n} \sum_{i=1}^{n} \left(Y_i - \widehat{Y}_i \right)^2 \tag{5}$$

where n is the size of the neural network batch, Y_i is the true value of the i th sample, and $\widehat{Y}_i$ is the predicted value of the i th sample. The smaller the value of MSE means the smaller the error between the true value and the predicted value, the better the model is fitted.

Fuzz Testing Evaluation Index. The failure discovery rate is an important evaluation index of fuzz testing, which is used to measure the ratio of the test cases generated by fuzz testing that can trigger the failure or crash of the UAV flight control system. The higher the failure discovery rate, the more effective the fuzz testing is in discovering the security holes or defects in the program. When using the configuration items constructed by the fuzz testing, according to the state of the drone defined in Sect. 3.6, record the results caused by each group of configuration items, and count the ratio of the configuration items that cause failures to the total test cases.

Coverage is the most important evaluation index of fuzz testing. Tests driven by coverage can quantify the progress of experiments and evaluate the pros and cons of fuzz testing methods. Code coverage includes indicators such as statement coverage, decision coverage, and branch coverage. We use statement coverage as the criterion for code coverage, that is, to detect whether all executable statements in the source code are executed. However, it is impossible to determine whether all logic functions have been tested with the help of code coverage alone, and the combination of function coverage can provide more accurate evaluation indicators for fuzz testing. The function coverage rate only counts the functions called. It ignores the statistics of the internal code of the function, which is used to detect whether the fuzz testing covers the functional characteristics required by the design specification. In the ArduPilot, the library that is strongly related to the flight state of the multi-rotor UAV is AC AttitudeControl, which includes the attitude and position control for the UAV. By adjusting the types and values of configuration items, the kinds of flight tasks, and the environment, we seek to improve the code coverage and function coverage of relevant source files in this code base to evaluate the adequacy of the fuzz testing method.

4.3 Analysis of Experimental Results

Analysis of Neural Network Results. The experiments compare the fitting ability of the LSTM and TCN used in LGDFuzz [3] and the TALA network model proposed in this paper to the UAV state under different flight missions. The specific performance difference results are shown in Table 2.

Table 2. Comparison of MSE of different models on test sets for different flight missions.

Flight Missions	MSE		
	LSTM	TCN	TALA
AVC2013	2.6307e−04	2.3051e−04	1.6591e−04
FitCollection	2.8905e−04	1.9475e−04	1.7528e−04
Copter-Mission	5.0408e−04	3.0376e−04	2.7091e−04
SDUT-Map	3.5327e−04	2.8599e−04	2.3811e−04

It can be seen from the table that the TALA neural network model has the smallest error value on the test set, which has certain advantages over other models. Specifically, in the AVC2013 mission, the MSE of TALA is reduced by 0.9716e−04 compared with the LSTM model; compared with the TCN model, the MSE is reduced by 0.6460e−04. For the SDUT-Map task with more complex flight paths, the MSE of TALA is reduced by 1.1516e−04 compared with the LSTM model; compared with the TCN model, the MSE is reduced by 0.4788e−04. The above results show that the TALA neural network model is effective for predicting UAV flight status, and the prediction accuracy is higher than other models. The TALA neural network can also maintain good generalization with lower error for different data sets.

Analysis of Fuzz Testing Results. For the failure discovery rate, we count the number and proportion of the UAV flight control system failures caused by the test cases generated by 4000 sets of fuzz testing in the sim_vehicle verification phase. The specific experimental results are shown in Table 3.

Table 3. Failure discovery rate.

UAV Status	Frequency	Proportion	Severity
Pass	361	9.0%	6
Timeout	814	20.4%	2
PreArm Failed	480	12.0%	0
Potential Thrust Loss	997	24.9%	0
Course Deviation	416	10.4%	4
Yaw Imbalance	367	9.2%	0
Spin	44	1.1%	0
Crash	521	13.0%	0

This table shows that among the test cases generated by the fuzz testing, the configuration items that can trigger the failure or crash of the UAV flight control system account for as high as 91%, and the proportion that can cause abnormal states after the UAV takes off reaches 88%. Only 361 groups of false positive configuration items can make the UAV complete the flight mission normally. The statistical results of the failure discovery rate fully prove the effectiveness of fuzz testing and provide support for the AutoTest verification phase.

In terms of coverage, we count the code coverage and function coverage of the AC_AttitudeControl.cpp, AC_AttitudeControl_Multi.cpp and AC_PosControl.cpp files in the AC_AttitudeControl pose control library. They represent the coverage levels of Universal Attitude Controller, Multi-rotor UAV Attitude Controller and Position Controller respectively. The comparison between our experimental results and the official results of ArduPilot is shown in Table 4.

Table 4. Coverage statistics for different files.

Controller Name	Code Coverage		Function Coverage	
	ArduPilot	Our Approach	ArduPilot	Our Approach
AC_AttitudeControl.cpp	75.8%	81.3%	89.2%	86.5%
AC_AttitudeControl_Multi.cpp	94.0%	97.7%	100%	100%
AC_PosControl.cpp	90.6%	91.3%	90.6%	90.6%

Among them, the code coverage and function coverage of the Universal Attitude Controller have reached more than 80%. Compared with the official coverage provided by ArduPilot [22], the code coverage has increased by 5.1%, but the function coverage has a difference of 2.7%. The code coverage rate of the human-machine attitude controller has risen by 3.7%, and the function coverage rate has reached 100%; the code coverage rate of the position controller has increased by 0.7%, and the function coverage rate is the same as the official one.

For files that do not reach 100% coverage first, to ensure that the ArduPilot can completely accept the test cases generated by the fuzz testing, all generated configuration items are within the security range officially defined by ArduPilot. This also resulted in some out-of-bounds checks and processing-related code not being executed. Second, part of the code is not subject to the configuration items involved in fuzz testing in this paper, so the relevant code lines are not covered. Finally, some redundant or unreachable codes are in the system but still participate in the coverage statistics. For the libraries that are not included in the coverage statistics, most of them are not related to the position and attitude of the drone; while the files that are not included in the coverage statistics in the library are mostly because the models are not related or the flight modes are not related. Complete code coverage testing is very difficult to achieve, and special test cases need to be manually designed, which cannot be achieved by fuzz testing some configuration items, and specific problems need to be analyzed in detail.

5 Conclusions

The tool performs fuzz testing based on the differential evolution algorithm for configuration items that affect the position and attitude of the UAV and performs coverage statistics on related code libraries to guide the selection of configuration items. In calculating the fitness function of the differential evolution algorithm, the TALA neural network is used to calculate the estimated value of the UAV state. The experimental results show that the TALA neural network can better extract the features of the flight logs and predict the state of the UAV with less error. Test cases generated by fuzz testing based on differential evolutionary algorithms are able to guide the UAV to seven abnormal states, in the verification stage. The coverage rate of the code library related to the position and attitude control of the UAV can exceed the data provided by the official ArduPilot, which proves the effectiveness of the method.

Acknowledgment. This study was supported by the National Key Research and Development Program of China (2020YFB1005704).

References

1. American Fuzzy Lop. https://lcamtuf.coredump.cx/afl/. Accessed 16 Dec 2022
2. Kim, T., et al.: RVFuzzer: finding input validation bugs in robotic vehicles through control-guided testing. In: 28th USENIX Security Symposium (USENIX Security 19). USENIX Association, Santa Clara, CA, pp. 425–442 (2019)
3. Han, R., et al.: Control parameters considered harmful: detecting range specification bugs in drone configuration modules via learning-guided search. In: 2022 IEEE/ACM 44th International Conference on Software Engineering (ICSE), Pittsburgh, PA, USA, pp. 462–473 (2022)
4. Petit, J., Stottelaar, B., Feiri, M., Kargl, F.: Remote attacks on automated vehicles sensors: experiments on camera and LiDAR. Black Hat Europe **11**, 995 (2015)
5. Fu, Z., Zhi, Y., Ji, S., Sun, X.: Remote attacks on drones vision sensors: an empirical study. IEEE Trans. Depend. Secure Comput. **19**, 3125–3135 (2022)
6. Yunmok, S., et al.: Rocking drones with intentional sound noise on gyroscopic sensors. In: Proceedings of the 24th USENIX Conference on Security Symposium, pp. 881–896 (2015)
7. Zou, Q., Huang, S., Lin, F., Cong, M.: Detection of GPS spoofing based on UAV model estimation. In: IECON 2016 - 42nd Annual Conference of the IEEE Industrial Electronics Society, Florence, pp. 6097–6102 (2016)
8. Bethi, P., Pathipati, S., Aparna, P.: Stealthy GPS spoofing: spoofer systems, spoofing techniques and strategies. In: 2020 IEEE 17th India Council International Conference (INDICON), New Delhi, India, pp. 1–7 (2020)
9. Young-Min, K., Jaemin, Y., Byeong-Moon, C., Yongsoon, E., Kyung-Joon, P.: Empirical analysis of MAVLink protocol vulnerability for attacking unmanned aerial vehicles. IEEE Access **6**, 43203–43212 (2018)
10. Honggfuzz. https://honggfuzz.dev/. Accessed 5 Jan 2023
11. The LLVM Compiler Infrastructure. https://llvm.org/. Accessed 10 Jan 2023
12. Kim, H., Ozmen, M.O., Bianchi, A., Celik, Z.B., Xu, D.: PGFUZZ: policy-guided fuzzing for robotic vehicles. In: NDSS (2021)

13. She, D., Pei, K., Epstein, D., Yang, J., Ray, B., Jana, S.: NEUZZ: efficient fuzzing with neural program smoothing. In: 2019 IEEE Symposium on Security and Privacy (SP), pp. 803–817 (2019)
14. Nichols, N., Raugas, M., Jasper, R.J., Hilliard, N.: Faster fuzzing: reinitialization with deep neural models. arXiv, abs/1711.02807 (2017)
15. Lv, C., et al.: SmartSeed: smart seed generation for efficient fuzzing. arXiv, abs/1807.02606 (2018)
16. Complete Parameter List. https://ardupilot.org/copter/docs/parameters-Copter-stable-V4.3.7.html. Accessed 16 June 2023
17. AVC 2013. https://avc.sparkfun.com/2013. Accessed 16 June 2023
18. Storn, R., Price, K.: Differential evolution: a simple and efficient heuristic for global optimization over continuous spaces. J. Glob. Optim. **11**(4), 341–359 (1997)
19. Geatpy: The genetic and evolutionary algorithm toolbox with high performance in python. http://www.geatpy.com/. Accessed 2 Mar 2023
20. Bai, S., Kolter, J.Z., Koltun, V.: An empirical evaluation of generic convolutional and recurrent networks for sequence modeling. arXiv preprint arXiv:1803.01271 (2018)
21. Luong, T., Pham, H., Manning, C.D.: Effective approaches to attention-based neural machine translation. In: Proceedings of the 2015 Conference on Empirical Methods in Natural Language Processing, Lisbon, Portugal, pp. 1412—1421 (2015)
22. ArduPilot LCOV - code coverage report. https://firmware.ardupilot.org/coverage/. Accessed 16 June 2023

A Survey of Algorithms for Addressing the Shortest Vector Problem (SVP)

Errui He[ID], Tianyu Xu[ID], Mengsi Wu[ID], Jiageng Chen[ID], Shixiong Yao[ID],
and Pei Li$^{(\boxtimes)}$[ID]

Central China Normal University, Wuhan 430079, Hubei, China
`peili@ccnu.edu.cn`

Abstract. Lattice-based encryption schemes derive their security from
the presumed complexity of solving the Shortest Vector Problem (SVP)
within the lattice structure. Numerous algorithms have been proposed to
address the SVP, targeting both exact and approximate solutions. However, it is noteworthy that the time complexity associated with these
algorithms predominantly exceeds polynomial bounds. This paper suc-
cinctly delineates these algorithms while providing a comprehensive summary of existing parallel implementations of sieving and enumeration
methods. Furthermore, it introduces distinguished instances from the
Hall of Fame of the SVP challenge.

Keywords: Lattice reduction · LLL · BKZ · Enumeration · Sieving ·
Parallel computing

1 Introduction

In recent years, with the rapid development of quantum computer technology,
the security of traditional cryptography has been threatened. Therefore, gov-
ernments and research institutions have begun to study secure cryptographic
algorithms under quantum computing models. Lattice-based cryptography has
obvious advantages in security, public-private key size, and resistance to quantum
attacks. Therefore, it is expected to replace existing cryptographic algorithms
and become a new cryptographic standard in the future.

In 1996, Ajtai gave a specification proof that the unique shortest vector problem (SVP) in lattices is under worst-case to that of the small integer solutions
(SIS) problem under average-case [1]. This proof can reduce the difficult problem in the lattice-based cryptography under worst-case to the a class of random
lattice problems, so the lattice-based cryptosystem can provide a worst-case
security proof.

A major factor in the security of lattice-based cryptographic protocols is the
difficulty of SVP in the lattice.This problem is crucial in lattice-based cryptography. There are many solutions to SVP problems, such as BKZ, enumeration, and

© ICST Institute for Computer Sciences, Social Informatics and Telecommunications Engineering 2024
Published by Springer Nature Switzerland AG 2024. All Rights Reserved
J. Chen and Z. Xia (Eds.), BlockTEA 2023, LNICST 577, pp. 57–76, 2024.
https://doi.org/10.1007/978-3-031-60037-1_4

sieving. In addition to being used separately, enumeration and sieving can also be embedded in the BKZ algorithm as sub-algorithms. The time complexity of the BKZ algorithm mainly depends on the time complexity of the sub-algorithm. The time complexity of the enumeration algorithm is $2^{\Omega(n)}$, and the time complexity of sieving algorithm is $2^{\Theta(n)}$, where n is the dimension of the lattice. These algorithms are super-polynomial time complexity, and how to accelerate the algorithm has become an important problem.

The enumeration is an exhaustive search algorithm used to find the shortest vector in a lattice. It achieves this by traversing all lattice points within an n-dimensional hypersphere centered at the origin with a radius of R. Practical enumeration can be viewed as traversing an enumeration tree, where each leaf node represents a lattice point. Pruning is a prevalent technique in enumeration, achieved by applying specific parameters to trim subtrees. While this approach introduces a notion of failure rate, it substantially reduces the search space, accelerating the algorithm's pace. Enumeration is often utilized as subroutines within BKZ, as they perform well in low dimensions but exhibit poor performance in high dimensions due to their super-exponential time complexity.

In 2001, [2] proposed the AKS Sieve, which was the first sieving algorithm. Apart from AKS Sieve, there is another sieving algorithm called the MV Sieve, which has a different structure. The ListSieve, introduced by [31], was the first sieving algorithm similar to MV. Both the AKS Sieve and the ListSieve are theoretical algorithms. Researchers subsequently developed heuristic versions based on these theoretical algorithms: NV Sieve [35] and GaussSieve [31].

Over the past decade, researchers have made various improvements to these sieving algorithms. The Level Sieve proposed by [41,45] takes the NV Sieve as the starting point. The Tuple Sieve proposed by [7] is based on the ListSieve. [34] proposed the Linear Sieve. The classic algorithm with the lowest complexity among heuristic algorithms is the LDSieve proposed by [8]. In addition, locality-sensitive hashing technology [23], locality-sensitive filtering technology [8] and quantum Grover search algorithm [25] are all used to speed up the traversal search process in the sieving algorithm. From the actual implementation effect level, the progressive sieving algorithm [24], sub-sieve algorithm [13] and G6K [3] optimize the actual execution effect of the sieving algorithm by using the rank-reduction technique. [27] also proposed a framework for the k-sieve algorithm with less memory usage. The sieving can also be used as a subroutine to improve the efficiency of BKZ [42].

This paper introduces the BKZ algorithm and summarizes the existing parallel implementations of enumeration and sieving. Table 1

2   Preliminaries

Notation. We let $\mathbb{Z}$ denotes the set of integers, $\mathbb{Q}$ denotes the set of rational numbers, and $\mathbb{R}$ denotes the set of real numbers. The nearest integer to a real number a, denoted by $\lceil a \rfloor$. For any real number a in the set $\mathbb{R}$, $|a|$ signifies the absolute value of a. It's noteworthy that all vectors are treated as column vectors.

Table 1. The parallel algorithm mentioned in this paper.

Algorithm	Author	Conference/Journal	Year	Type
HSB+10 [19]	Hermans et al.	AFRICACRYPT	2010	ENUM
DS10 [12]	Dagdelen and Schneider	Euro-Par	2010	ENUM
KSD+11 [22]	Kuo et al.	CHES	2011	ENUM
CMP16 [11]	Correia et al.	PDP	2016	ENUM
p3Enum [10]	Burger et al.	ICCS	2019	ENUM
p3EnumOpt [9]	Burger et al.	HPCS	2019	ENUM
MAP-SVP [39]	Tateiwa et al.	SC	2020	ENUM
EBD+21 [15]	Esseissah et al.	Scientific Programming	2021	ENUM
MS11 [33]	Milde et al.	PaCT	2011	Sieving
IKM+14 [20]	Ishiguro et al.	PKC	2014	Sieving
MTB14 [30]	Mariano et al.	Euro-Par	2014	Sieving
MBL15 [28]	Mariano et al.	ICPP	2015	Sieving
MLB17 [29]	Mariano et al.	PDP	2017	Sieving
AG20 [4]	Andrzejczak et al.	ICA3PP	2020	Sieving
DSW+21 [14]	Ducas et al.	EUROCRYPT	2021	Sieving

The Euclidean norm of a vector $\mathbf{a}$ belonging to the space $\mathbb{R}^m$ is represented as $\|\mathbf{a}\|$. Additionally, the inner product of vectors $\mathbf{a}$ and $\mathbf{b}$, both belonging to the space $\mathbb{R}^m$, is represented by $\langle \mathbf{a}, \mathbf{b} \rangle$.

Lattice. A lattice $\mathcal{L} = \{\mathbf{Bx} : \mathbf{x} \in \mathbb{Z}^n\}$ is denoted as $\mathcal{L}(\mathbf{B})$, where $\mathbf{B} = \{\mathbf{b}_1, \mathbf{b}_2, \ldots, \mathbf{b}_n\} \in \mathbb{R}^{m*n}$ is an ordered group of linear independent vectors, and $\mathbf{B}$ is called the basis of the lattice.

Gram-Schmidt Orthogonalisation. For a linearly independent ordered vector group $\mathbf{B} = \{\mathbf{b}_1, \mathbf{b}_2, \ldots, \mathbf{b}_n\} \in \mathbb{R}^{m*n}$, its Gram-Schmidt Orthogonalisation (GSO) set $\mathbf{B}^* = \{\mathbf{b}_1^*, \mathbf{b}_2^*, \ldots, \mathbf{b}_n^*\}$ is given by the following definition.

- $\mathbf{b}_1^* = \mathbf{b}_1$
- for $i > 1$, $\mathbf{b}_i^* = \mathbf{b}_i - \sum_{j=1}^{i-1} \mu_{i,j} \mathbf{b}_j^*$

where, for $1 \le i \le j \le n$, the GSO coefficient is calculated by the following formula:

$$\mu_{i,j} = \frac{\langle \mathbf{b}_i, \mathbf{b}_j^* \rangle}{\langle \mathbf{b}_j^*, \mathbf{b}_j^* \rangle}$$

Obviously, for all $1 \le i \le n$, $\mu_{i,i} = 1$.

Orthogonal Projections. For a vector $\mathbf{v} \in \mathcal{L}(B)$, its projections $\pi_i(\mathbf{v})$ are defined for $1 \le i \le n$ as follows:

- $\pi_1(\mathbf{v}) = \mathbf{v}$
- for $2 \le i \le n$, $\pi_i(\mathbf{v})$ is the orthogonal projection of $\mathbf{v}$ to $\text{span}(\mathbf{b}_1, \mathbf{b}_2, \ldots, \mathbf{b}_{i-1})^{\perp}$.

The projection $\pi_1(\mathbf{v})$ is written as follows:

$$\pi_i(\mathbf{v}) = \mathbf{v} - \sum_{j=1}^{i-1} \frac{\langle \mathbf{v}, \mathbf{b}_j^* \rangle}{\langle \mathbf{b}_j^*, \mathbf{b}_j^* \rangle}$$

In particular, when the vector $\mathbf{v}$ is the basis vector $\mathbf{b}_k$ of the lattice, it can be written as follows:

$$\pi_i(\mathbf{b}_k) = \mathbf{b}_k - \sum_{j=1}^{i-1} \mu_{k,j} \mathbf{b}_j^* = \mathbf{b}_k^* + \sum_{j=i}^{k-1} \mu_{k,j} \mathbf{b}_j^*$$

In the simplest scenario, $\pi_i(\mathbf{b}_i) = \mathbf{b}_i^*$.

Shortest Vector Problem (SVP). Let $\lambda_1, \ldots, \lambda_n$ denote the successive minima of lattice $\mathcal{L}$, $\lambda_i = \lambda_i(\mathcal{L})$ is defined as the smallest radius r of a ball that is centred at the origin and which contains r linearly independent lattice vectors.

Definition 1. Shortest Vector Problem (SVP). Given a lattice basis $\mathbf{B}$, the task is to find the shortest non-zero lattice vector, denoted as $\mathbf{v} \in \mathcal{L}(\mathbf{B})$, such that its norm satisfies $\|v\| = \lambda_1(\mathcal{L}(\mathbf{B}))$.

Ordering of Basis Vectors. Consider S_n as the group of permutations of elements in $\{1, 2, \ldots, n\}$. For $\sigma \in S_n$, define $\sigma(\mathbf{B}) = \{\mathbf{b}_{\sigma(1)}, \mathbf{b}_{\sigma(2)}, \ldots, \mathbf{b}_{\sigma(n)}\}$ as a vector ordering of $\mathbf{B}$. In particular, for $1 \leq i < k \leq n$, we define the following operation

$$\sigma_{i,k}(j) = \begin{cases} j, & j < i \text{ or } j > k \\ k, & j = i \\ j-1, & i+1 \leq j \leq k \end{cases}$$

For lattice basis $\mathbf{B}$, the operation is as follows:

$$\sigma_{i,k}(\mathbf{B}) = \{\mathbf{b}_1, \ldots, \mathbf{b}_{i-1}, \mathbf{b}_k, \mathbf{b}_i, \ldots, \mathbf{b}_{k-1}, \mathbf{b}_{k+1}, \ldots, \mathbf{b}_n\}$$

where $\mathbf{b}_k$ is inserted between $\mathbf{b}_{i-1}$ and $\mathbf{b}_i$, $\mathbf{b}_i, \ldots, \mathbf{b}_{k-1}$ moved one bit back.

3 Blockwise Korkine Zolotarev (BKZ) Reduction

Given the lattice basis $\mathbf{B}$, we can calculate its GSO matrix $\mathbf{B}^*$ and its GSO coefficients $\mu_{i,j}$ in polynomial time.

3.1 Size Reduction

Definition 2. (Size-reduced basis) A basis $\mathbf{B} = \{\mathbf{b}_1, \mathbf{b}_2, \ldots, \mathbf{b}_n\} \in \mathbb{R}^{m*n}$ is considered size reduced if for all $1 \leq j < i \leq n$, $|\mu_{i,j}| \leq \frac{1}{2}$.

Algorithm 1: Size-reduction algorithm for k-column vector of basis **B**

Input: A basis **B** along with its Gram-Schmidt orthogonalization (GSO) coefficients $\mu_{i,j}$, and an index k.

Output: A basis $\mathbf{B}'$ where the kth column vector $\mathbf{b}'_k$ satisfies size-ruduced, and the updated coefficients $\mu_{k,j}$

1: **for** $j = k - 1, \ldots, 1$ **do**
2:    **if** $|\mu_{k,j}| > \frac{1}{2}$ **then**
3:       $\mathbf{b}_k \leftarrow \mathbf{b}_k - \lceil \mu_{k,j} \rfloor \mathbf{b}_i$
4:       **for** $i = 1, \ldots, j$ **do**
5:          $\mu_{k,i} \leftarrow \mu_{k,i} - \lceil \mu_{k,j} \rfloor \mu_{j,i}$
   return Basis $\mathbf{B}'$ where the kth column vector $\mathbf{b}'_k$ satisfies size-ruduced, and the updated coefficients $\mu_{k,j}$.

Algorithm 1 gives the size-reduced operation. When we set $\mu_{k,j} \leftarrow \mu_{k,j} - \lceil \mu_{k,j} \rfloor \mu_{j,j}$, we get $|\mu_{k,j}| \leq \frac{1}{2}$ ($\mu_{j,j} = 1$). For consistency, we need to update the value of $\mathbf{b}_k$.

$$\mu'_{k,j} = \mu_{k,j} - \lceil \mu_{k,j} \rfloor \mu_{j,j} = \frac{\langle \mathbf{b}_k, \mathbf{b}_j^* \rangle}{\langle \mathbf{b}_j^*, \mathbf{b}_j^* \rangle} - \lceil \mu_{k,j} \rfloor \frac{\langle \mathbf{b}_j, \mathbf{b}_j^* \rangle}{\langle \mathbf{b}_j^*, \mathbf{b}_j^* \rangle} = \frac{\langle \mathbf{b}_k - \lceil \mu_{k,j} \rfloor \mathbf{b}_j, \mathbf{b}_j^* \rangle}{\langle \mathbf{b}_j^*, \mathbf{b}_j^* \rangle}$$

Then we set $\mathbf{b}'_k \leftarrow \mathbf{b}_k - \lceil \mu_{k,j} \rfloor \mathbf{b}_j$. Because we changed the value of $\mathbf{b}_k$, all the values of $\mu_{k,i} (1 \leq i < j)$ need to change with it.

$$\mu'_{k,i} = \frac{\langle \mathbf{b}_k - \lceil \mu_{k,j} \rfloor \mathbf{b}_j, \mathbf{b}_i^* \rangle}{\langle \mathbf{b}_i^*, \mathbf{b}_i^* \rangle} = \frac{\langle \mathbf{b}_k, \mathbf{b}_i^* \rangle}{\langle \mathbf{b}_i^*, \mathbf{b}_i^* \rangle} - \lceil \mu_{k,j} \rfloor \frac{\langle \mathbf{b}_j, \mathbf{b}_i^* \rangle}{\langle \mathbf{b}_i^*, \mathbf{b}_i^* \rangle} = \mu_{k,i} - \lceil \mu_{k,j} \rfloor \mu_{j,i}$$

For $1 \leq i < j < k$, we set $\mu_{k,i} \leftarrow \mu_{k,i} - \lceil \mu_{k,j} \rfloor \mu_{j,i}$. Obviously, when upon a size reduction of $\mathbf{b}_k$ with $\mathbf{b}_j$, we also change the value of $\mu_{k,i}$ ($1 \leq i < j$). Therefore, the value of j must be traversed from $k - 1$ to 1.

3.2 LLL Reduction

Definition 3. ($\delta - $ **LLL reduced basis**) Given $\frac{1}{4} < \delta \leq 1$, a basis $\mathbf{B} = \{\mathbf{b}_1, \mathbf{b}_2, \ldots, \mathbf{b}_n\} \in \mathbb{R}^{m*n}$ is considered $\delta - LLL$ reduced if the following conditions are met.

- **B** is size reduced as defined in Definition 2.
- For all $2 \leq k \leq n$, $\delta \|\pi_{k-1}(\mathbf{b}_{k-1})\|^2 \leq \|\pi_{k-1}(\mathbf{b}_k)\|^2$.

Algorithm 2: δ-LLL-reduction algorithm, described in [26]

Input: A basis $\mathbf{B}$ and $\frac{1}{4} < \delta \leq 1$.

Output: A basis $\mathbf{B}'$ which is δ-LLL reduced

1: Find the GSO basis $\mathbf{B}^*$ and GSO coefficients $\mu_{i,j}$
2: $k \leftarrow 2$
3: **while** $k \leq n$ **do**
4:     Size-reduce $\mathbf{b}_k$ ▷ *Use Algorithm 1*
5:     **if** $\|(\mathbf{b}_k^*)\|^2 < (\delta - \mu_{k,k-1}^2)\|(\mathbf{b}_{k-1})\|^2$ **then**
6:         $\mathbf{B} \leftarrow \sigma_{k-1,k}(\mathbf{B})$
7:         Update $\mathbf{b}_{k-1}^*, \mathbf{b}_k^*$ and $\mu_{i,j}\mathrm{s}(i \geq k-1)$.
8:         $k \leftarrow \max(k-1, 2)$
9:     **else**
10:        $k \leftarrow k+1$
    return A basis $\mathbf{B}'$ which is δ-LLL reduced.

As can be seen from the content of Sect. 2, $\|\pi_{k-1}(\mathbf{b}_{k-1})\|^2 = \|(\mathbf{b}_{k-1}^*)\|^2$, and $\|\pi_{k-1}(\mathbf{b}_k)\|^2$ can be written as $\|\mathbf{b}_k^*\|^2 - \mu_{k,k-1}^2\|\mathbf{b}_{k-1}^*\|^2$. Therefore, the condition of δ-LLL reduction can be written equivalently as $(\delta - \mu_{k,k-1}^2)\|(\mathbf{b}_{k-1})\| \leq \|(\mathbf{b}_k^*)\|^2$. Based on this condition, we get Algorithm 2 of LLL-reduced.

Algorithm 3 that describes BKZ consists of a LLL algorithm and a sub-algorithm that accurately solves the SVP algorithm. Its specific definitions and algorithms are as follows.

Definition 4. (**δ-Korkine Zolotarev reduced basis**) Given $\frac{1}{4} < \delta \leq 1$, a basis $\mathbf{B} = \{\mathbf{b}_1, \mathbf{b}_2, \ldots, \mathbf{b}_n\} \in \mathbb{R}^{m*n}$ is said to be δ-Korkine Zolotarev reduced if the following conditions are satisfied.

- The basis $\mathbf{B}$ is size-reduced, as defined in Definition 1.
- For all $1 \leq i \leq n$, $\delta\|\mathbf{b}_i\| = \lambda_1(\pi_i(\mathcal{L}))$.

In Algorithm 3, we use *subalgorithm* to represent the exact solution algorithm for SVP.

Algorithm 3: δ-BKZ reduction algorithm

Input: A basis $\mathbf{B}$, An integer β and $\frac{1}{4} < \delta \leq 1$.

Output: A basis $\mathbf{B}'$ which is δ-BKZ reduced

1: Find the GSO basis $\mathbf{B}^*$ and GSO coefficients $\mu_{i,j}$
2: $i \leftarrow 1$
3: LLL reduce $\mathbf{B}$
4: **while** $i \leq n$ **do**
5:     $k \leftarrow \min(i + \beta - 1, n)$
6:     *subalgorithm* $(\{\mathbf{b}_i, \ldots, \mathbf{b}_k\})$
7:     $T \leftarrow \|\mathbf{b}_i\|$
8:     LLL reduce $\mathbf{b}_i, \ldots, \mathbf{b}_n$
9:     **if** $T = \|\mathbf{b}_i\|$ **then**
10:        $i \leftarrow i+1$
    return A basis $\mathbf{B}'$ which is δ-BKZ reduced.

The sub-algorithms of the sixth line of the algorithm are mostly enumerations and sieving, and the time complexity of these algorithms is exponential. So we divide the entire lattice base into blocks of size b and run sub-algorithms on each block. In the seventh line, we record the length of the first vector in the block and LLL reduce the subsequent vectors. If the length of the vector does not change after LLL reduce, the index value is incremented by 1.

4   Enumeration

This section introduces various variants of enumeration algorithms along with their corresponding developments in parallel implementation.

4.1   The Basic Schnorr-Euchner Enumeration

While the earliest enumeration algorithm can be traced back to the 1980 s [16,21, 36], the first practical enumeration algorithm was introduced in 1994 by Schnorr and Euchner [37] as a subroutine of BKZ algorithm. This algorithm is referred to as the basic Schnorr-Euchner enumeration in [44], which treats the enumeration process as a traversal of the enumeration tree using depth-first search. During the enumeration process, all leaf nodes, representing lattice points, are traversed. Its practicality lies in its earliest introduction of pruning method. Pruning involves setting boundary functions to ensure that different layers of the enumeration tree have corresponding boundary values. If the target value generated by a node in a certain layer does not meet the required boundary value, the subtree corresponding to that node is discarded. The comprehensive description of this algorithm can be found in [37,44].

The enumeration algorithm in [37], as the first practical enumeration algorithm, serves as the foundation for subsequent algorithm improvements. Inspired by this enumeration algorithm, the following three notably significant parallel implementations were introduced in the subsequent years [12,15,19]:

HSB+10 [19]: The earliest application of GPU for accelerating the solution of the SVP using the enumeration algorithm was conducted by Hermans et al. in 2009. This parallel implementation based on [37]. The main idea was to split the enumeration tree at a higher level and then pass the split subtrees to the GPU for enumeration. This splitting approach eliminated communication between subtrees. However, due to the uncertain initial subtree sizes, termination times among threads varied. To tackle this, the authors introduced the Early Termination for improvement. The authors used the NVIDIA GTX 280 GPU, which provided nearly a 5-fold speedup compared to the serial enumeration algorithm in the fplll. However, the article only utilized a single CPU core and did not fully exploit the computational power of the CPU during the subtree enumeration phase. Additionally, more advanced pruning methods were not used.

DS10 [12]: In 2010, Dagdelen et al. introduced a parallelized version of the basic Schnorr-Euchner enumeration [37] based on multi-core CPUs. The main idea was to divide the entire enumeration tree into different subtrees, which were placed in a shared subtree queue. Whenever a thread became idle, it could retrieve an unenumerated subtree from the subtree queue for enumeration. Each thread that obtained a new subtree was only allowed to partition it once. However, this approach introduced the problem of imbalanced subtree sizes because the subdivision of subtrees was not based on the size of the tree. The authors addressed this issue by considering the depth of the nodes. The authors conducted comparative experiments using different numbers of CPU cores. This parallel implementation achieved superlinear speedup, which was attributed to the shared minimum value A among different threads. This introduced some communication overhead but also resulted in better pruning effects. Although the single-threaded program in the article was slower than the enumeration algorithm provided by fplll, significant acceleration was observed when utilizing multiple cores in parallel. One limitation of this study was that only a single CPU was used, and future research should investigate how to leverage multi-core and multi-node CPUs to accelerate enumeration algorithms.

EBD+21 [15]: In 2021, Esseissah et al. made improvements to [19] using three strategies. For the preprocessing part, they employed a combination of randomization and Gaussian heuristics to obtain better bases. Concerning communication, [19] required CPU-to-GPU data transfer for each enumeration, leading to significant overhead. The authors proposed generating a portion of lattice points on the GPU to reduce communication load and enhance efficiency. The experiments were conducted using NVIDIA GeForce GTX 1060 GPU. Compared to [19], the proposed method achieved over 2.5 times speedup in a 110-dimensional lattice. Using two GPUs resulted in nearly linear or even superlinear speedup compared to using a single CPU. This indicates the scalability of the algorithm. The performance using two GPUs is superior to Correia's implementation [11] using a 16-core CPU in a 60-dimensional lattice. However, the article also has some limitations, such as not utilizing the latest GPU architecture and only conducting experiments with two GPUs. In the future, further research could explore large-scale experiments with multiple GPUs.

4.2 Extreme Pruning

Since 1994, when pruning methods were introduced into enumeration by Schnorr and Euchner [37], pruning has undergone several developments, solidifying its key role in enumeration. In 1995, Schnorr and Hörner [38] introduced an alternative pruning method based on the Gaussian volume heuristic. Herman et al. [19] also mentioned their intention to integrate this pruning approach into future implementations. However, both of these pruning methods lacked detailed analysis. It wasn't until 2010 that Gama et al. [17] presented a thorough theoretical analysis of diverse pruning approaches and introduced the concept of extreme pruning. Extreme pruning outperformed previous pruning-based enumeration algorithms in practice. In 2018, Aono et al. [5] demonstrated the lower bound of the cost of extreme pruning technology, further solidifying its effectiveness.

Starting from 2010, the parallel implementation of extreme pruning method has also garnered widespread attention from researchers [9, 10, 22]:

KSD+11 [22]: In 2011, Kuo et al. proposed a parallel enumeration algorithm using extreme pruning. Due to the introduction of the failure rate in the extreme pruning, in this approach, it was necessary to enumerate different enumeration trees corresponding to different bases of the same lattice. The SVP solver combines MapReduce technology. The GPU and MapReduce implementations of this algorithm were primarily based on [19], which employed a polynomial pruning function. In the experiment, the authors resolved the 114-dimensional SVP Challenge in approximately 40 h on a single workstation with eight NVIDIA GeForce GTX 480 GPUs. Additionally, through integrating recent techniques and leasing an Amazon server for \$2,300, they successfully addressed the 120-dimensional SVP Challenge. However, the authors also acknowledged the difficulty in finding a good pruning function.

p3Enum [10]: In 2019, Burger et al. proposed p3Enum, an open-source framework tailored for parallelizing enumeration with extreme pruning, specifically devised for addressing the SVP. This parallel implementation was inspired by [6] and [22]. This paper introduced a new parameter, v, to the pruning function. This parameter increased the probability of successful enumeration and the workload, allowing for higher success rates when running for the same amount of time as the single-threaded program. The authors conducted experimental comparisons with well-known algorithms at the time and found that p3Enum was the fastest SVP solver in the 66–88 dimensional range. And p3Enum surpasses the GPU enumeration from [19] within dimensions 76–90. Moreover, on a 60-core system, they achieved a parallel efficiency surpassing 0.9. The authors intended to enhance the performance of p3Enum in their future work by incorporating MPI technology.

p3EnumOpt [9]: In 2019, Burger et al. made two improvements to the SVP solver p3Enum, called p3EnumOpt. First, they proposed a novel parallel implementation for extreme pruning. Second, they introduced the p3Enum extreme pruning function generator (p3Enum-epfg), which utilizes a parallelized simulated annealing approach to generate optimized extreme pruning functions for p3Enum's pruned enumeration. The authors conducted experiments to compare the efficiency of this new algorithm with p3Enum [10], fplll, and SubSieve. They found that the new algorithm had the fastest execution speed for solving SVP in the 66 to 92-dimensional range. However, the serial version of this algorithm was slower than fplll, and there was a lack of exploration for SVP in higher-dimensional lattices. This presents a potential direction for future research.

MAP-SVP [39]: In 2020, Tateiwa et al. introduced the world's inaugural distributed and asynchronous parallel SVP solver, referred to as the MAssively Parallel solver for SVP (MAP-SVP). The system is made up of a lot of Solvers and a management procedure named LoadCoordinator. In addition to running the DeepBKZ and ENUM algorithms, each Solver also exchanges short vectors through a vector pool that is under the control of a load coordinator. Extreme pruning procedures are used by every Solver. More aggressive pruning is used to shorten the computation time for each Solver as the number of Solvers grows. Using 100,032 cores, the authors carried out the SVP research's largest-scale studies. In addressing instances of the SVP Challenge, they achieved new marks for dimensions 104, 111, 121, and 127 using this concurrent program. On the Fugaku supercomputer at RIKEN, the authors also want to carry out much larger-scale SVP-solving experiments involving millions of processes.

4.3   The Closest Point Search Algorithm: SE++

SE++ is a variant of an enumeration algorithm proposed by Ghasemmehdi and Agrell [18]. Experiments indicate that on a 60-dimensional lattice, it can reduce 75% of floating-point computations. The parallel implementation of this method was studied by Correia et al. [11]:

CMP16   [11]: In 2016, Correia et al. introduced a parallelized version of the practical enumeration algorithm SE++ [18] for solving the CVP, which can also be used to solve the SVP. This parallel implementation involved processing different branches of the tree in parallel and performing separate enumerations for each branch. Additionally, it avoided redundant calculations for symmetric branches in the tree, thus improving efficiency. The authors conducted experiments and comparisons, and found that SE++ outperformed the algorithm in [12] by 35% to 60% in terms of speed.

4.4   Comparison

In a broader perspective, the advancements in GPU and multi-core CPU technologies have propelled the evolution of parallel implementation within enumeration algorithms, significantly expediting the resolution speed of the Shortest Vector Problem (SVP). According to Table 2, the most efficient algorithm for lower dimensions (66–92) is currently p3EnumOpt [9]. Moreover, the MAP-SVP [39], utilizing extensive parallelism, has effectively addressed the 127-dimensional SVP Challenge. Both of these algorithms rely on extreme pruning, underscoring the pivotal role of this technique in parallel implementation. Nevertheless, up to the present point, the record holder for the SVP Challenge remains sieve, achieving a solution for the 186-dimensional SVP Challenge. This reality signifies the substantial scope for improvement within enumeration algorithms.

5   Sieving

The core concept of the sieving algorithm involves creating a list, denoted as L, comprising N lattice vectors. Subsequently, the algorithm engages in pairwise reduction, systematically replacing the longest vector within the list. The process concludes when the list saturates a ball with a radius R. In other words, the algorithm terminates when L encompasses a substantial portion of short vectors, each having a length no greater than R. Algorithm 4 provides a concise summary of this straightforward sieving algorithm.

Table 2. Performance comparison of parallel enumeration algorithms.

Year	Algorithm	Based	Device	Performance
2010	HSB+10 [19]	SE94 [37]	Tesla GPU	5x fplll-3.0.11 (dim=50–56, pre-reduction=LLL) (dim=58, pre-reduction=BKZ-20)
2010	DS10 [19]	SE94 [37]	16 Cores CPU	14x single-core (dim=52) 11–12x fplll-3.0.12 (dim=52)
2011	KSD+11 [22]	GNR10 [17]	Fermi GPU	New Records in 114, 116, 120
2016	CMP16 [11]	SE++ [18]	16 Cores CPU	1.35–1.6x DS10 [19] (dim=40, 50, 60)
2019	p3Enum [10]	GNR10 [17]	multi-core CPU	Best in 66–88
2019	p3EnumOpt [9]	GNR10 [17]	multi-core CPU	Best in 66–92
2020	MAP-SVP [39]	GNR10 [17]	100,032 cores	New Records in 104, 111, 121, 127
2021	EBD+21 [15]	SE94 [37]	Pascal GPU	1 GPU: 2.5x HSB+10 [19] 2 GPU: 5x HSB+10 [19] (dim=100, 110)

With the development of multi-core CPUs, GPUs, and FPGAs, many software developers and researchers are adopting parallel programming techniques to take advantage of these hardware resources. The sieving is no exception. In 2011, [33] proposed a parallel GaussSieve algorithm, but once the number of threads exceeds 10, the acceleration factor will not exceed about 5. So, [20] proposed a more practical parallel GaussSieve algorithm based on this algorithm, and [43] also used this framework to implement it on GPUs. Then, [30] proposed a parallel lock-free GaussSieve using a shared linked list. And then [28] also proposed a parallel lock-free HashSieve. With the proposal of LDSieve, [29] also used a similar lock-free mechanism to implement parallel LDSieve. Subsequently, [4] proposed a multiplatform parallel approach using FPGAs. Finally, [14] proposed two parallel bucketing methods, BGJ-like bucketing and BDGL-like bucketing, and a parallel reducing method using Tensor cores.

These algorithms can be roughly divided into three types of parallel optimization algorithms, parallel GaussSieve, parallel HashSieve and parallel LDSieve algorithms. The performance of these algorithms is contrasted in Table 3.

Algorithm 4: Sieving algorithm, described in [14]

Input: A basis $\mathbf{B} = (\mathbf{b}_1, ..., \mathbf{b}_n)$ of a lattice $\mathcal{L}$, a saturation radius R, and a list L
Output: A list L comprising short vectors that saturate the ball of radius R
1: Sample a lot of vectors from the lattice $\mathcal{L}$ in L.
2: **while** L does not cover the entire ball with a radius of R **do**
3: Find reduction and replace the longest vector in L.
   return L

Table 3. The Performance, TC (Time Complexity) and SC (Space Complexity) comparison of each algorithm.

Year	Algorithm	Service	Performance	TC	SC
2011	MS11 [33]	CPU	thread$\leq$ 5, linear growth	-	-
2014	IKM+14 [20]	CPU	using 29,994 CPU hours in 128	$2^{0.52n}$	$<2^{0.2n}$
2014	MTB14 [30]	CPU	1.12x MS11 [33], 1.50x IKM+14 [20]	-	-
2020	AG20 [4]	FPGA	8x and only 6% CPU overhead	-	-
2015	MBL15 [28]	CPU	2.5x GuassSieve	$2^{(0.32n-15)}$ ~ $2^{(0.33n-16)}$	-
2017	MLB17 [29]	CPU	50x over the original LDSieve	-	-
2021	DSW+21 [14]	GPU	New Record in 180	$2^{0.367n}$ and $2^{0.338n}$	$2^{0.2075n}$

5.1 Parallel Gauss Sieve Algorithms

The main idea of the GaussSieve algorithm is to sample lots of vectors through a sampling algorithm. First, the algorithm uses the vectors in the list to reduce the sampling vectors, and then uses the reduced sampling vectors to reduce the vectors in the list. If reduction occurs, the sampling vectors will replace the corresponding vectors in the list.

The initial parallel implementation of the GaussSieve was introduced in 2011 by [33]. This algorithm aims to create a parallelized version version of the original GaussSieve algorithm. In this algorithm, let t represent the number of threads. Every thread possesses its unique instance and independently performs the sieving algorithm. These instances are ring-connected to one another. If the vector $\mathbf{v}$ is unchanged after reduction steps, it will be transmitted to the buffer Q_{i+1} in the subsequent instance. On the contrary, if $\mathbf{v}$ has any changes, it will be transferred to the distributed stack S_i within its respective instance. If a vector $\mathbf{v}$ successfully traverses all instances, it will be appended to the distributed list L_i. For a small number of up to 5 parallel threads, the scalability of the parallel version is nearly linear. Nevertheless, the algorithm fails to guarantee the sustained pairwise reduction of the entire list L. This is attributed to the fact that, upon the addition of a vector to the distributed list, another instance might introduce additional vectors. Consequently, the overall performance of this algorithm does not experience acceleration, particularly when dealing with a substantial number of threads.

In 2014, an alternative parallel version of GaussSieve was introduced by [20], designed for both shared and distributed memory systems, exhibiting enhanced scalability, especially up to 8 threads. The implementation capitalizes on the property of pairwise-reduced sets that have been unified, which is as follows: If two sets, A and B, have undergone pairwise reduction, and every vector pair ($\mathbf{a}$, $\mathbf{b}$) is Gauss-reduced for $\forall \mathbf{a} \in A$, $\mathbf{b} \in B$, then the union $A \cup B$ remains pairwise-reduced. The algorithm takes a list V comprising r samples as input and follows a three-step reduction process.

During the initial phase, the sample vectors in V undergo simultaneous reduction with respect to the vectors in L, following the approach of the initial GaussSieve algorithm. Throughout this process, each modified vector is incorporated into the stack S, while unaltered vectors are transferred to the list V'. In the subsequent step, the initial vectors in V' undergo parallel reduction with each other. Modified vectors are relocated to the stack S, whereas unaltered vectors are shifted to the list V''. In the last step, each thread executes reduction operations on a portion of L against the vectors in V''. Similarly, if any vector changes, it is appended to the stack S; Otherwise, it is included in L'. Much like V'', L' also maintains pairwise reduction.

Upon completion of each iteration, the lists L' and V'' are combined to form the new list L, and the vectors in S populate the list V. This entire process continues until the count of collisions, denoted as K, reaches a predetermined threshold value, denoted as c. So this algorithm can solve the problem that there are non-Gauss-reduced pairs in the reduced list. However, this algorithm has two disadvantages: utilizing r samples elevates the computational load, and determining the optimal value for the parameter r in advance is not feasible. Additionally, insertions in S occur sequentially, posing limitations on scalability.

Therefore, [30] proposed a GuassSieve shared memory parallel algorithm based on a lock-free linked list. They have better scalability than [20] and better performance than the results reported in [33]. [30] used the Harris linked list to store vectors according to the length of the inner product. To avoid two threads modifying the same vector at the same time, a thread intending to modify the vector should first retrieve it from the list, and subsequently, insert the modified version. To achieve this, they made slight adjustments to the *Reduce* function in the gsieve library. The *Reduce* function was divided into two separate functions, namely *TestReduce* and *eReduce*. *TestReduce* evaluates whether **v** should undergo reduction against **w**. If the outcome is true, the vector **w** is taken out from L and then duplicated into another variable. The copy of **w** is subsequently modified using *eReduce*. If the outcome is false, the *eReduce* operation is not invoked. Thus the algorithm ensures that the vector is never modified in L. The entire algorithm process is as follows : first, all threads perform the reduction of **v** by **w**, and **w** $\in L$. Subsequently, vector **v** is scheduled for insertion into set L in a manner that ensures L maintains its order.

This algorithm may have the same problem as [33], but they relaxed the condition. While a vector **v** may not be reduced relative to another vector **w**, the vectors in their reduced forms are likely to undergo further reduction in relation to each other. In addition, two optimizations of the algorithm were also suggested by [30]: the samples utilized in the sieving process are of shorter length, and the reduction process predominantly targets the shorter vectors themselves. The final achieved performance and scalability are much better than [33], and outperforms [20], with almost 1.12x and 1.50x performance improvement.

Finally, in 2020, [4] proposed an accelerator to accelerate the GuassSieve algorithm, and gave a multiplatform parallel implementation approach. The accelerator mainly accelerates two parts of the Reduce algorithm: dot product calculation and the modification of the vector's value. In the computation of the dot product, the initial step entails multiplication with the respective coefficients. Subsequently, the result is directed to an additive tree composed of $\lceil log_2(n)/\beta \rceil$ additive layers. Here, β represents the number of additions executed in a single clock cycle, with shorter delay paths compared to multiplications. Then, the clock delay for the dot product calculation is $1+\lceil log_2(n)/\beta \rceil$. After the dot product calculation, the algorithm need compute $q = \lceil dot/||\mathbf{u}||^2 \rceil$. This operation does not perform division, but rather checks conditions. This operation requires only one clock cycle. For the update of the vector's value $||\mathbf{v}||^2+ = q^2 \cdot ||\mathbf{u}||^2 - 2 \cdot q \cdot dot$, the hardware performs the norm update function in three sequential steps, with each step consuming one clock cycle. In the first step, the hardware calculates q is multiplied with the two interim results. Subsequently, the subtraction and addition operations are conducted. Finally, the delay of reducing a pair of vector depends solely on the dimensions of the lattice. In the case of an n-dimensional lattice, the delay $f_{cl}(n)$ is exactly equal to $f_{cl}(n) = \lceil log_2(n)/\beta \rceil + 5$.

A multi-platform approach to caching can eliminate data transfer bottlenecks. This algorithm is divided into three steps. The initial step involves reducing the set S by utilizing the reduced vectors from the list L. Let the size of S be $k = f_{tl}(n,w)$. The data transfer delay for a vector, denoted as $k = f_{tl}(n,w) = \lceil (n+2) \cdot 16/w \rceil$. First, this approach transfers the k vectors of the set S from the CPU to the FPGA, and a vector of the list L, and then begins to reduce. The next vector of the list L is transmitted while reducing. The delay is $f_{el}^{p1}(n,w) = k^2 + k \cdot |L| + f_{cl}(n)$. During the second phase, the vectors in the set S are self-reduced, and as S is already resident in the FPGA, communication overhead is effectively eliminated. Hence, the delay is $f_{el}^{p2}(n,w) = k^2/2 \cdot f_{cl}(n) + k^2/2 + f_{cl}(n)$. In the final phase, the vectors in list L undergo reduction by the vectors in set S, akin to the first step. So the delay is $f_{el}^{p3}(n,w) = k \cdot |L| + f_{cl}(n)$. For the specified 160-dimensional lattice, the suggested resolution is anticipated to yield a performance improvement of 8.32 times compared to CPU. And the proposed architecture requires only 6% of the CPU-based cost.

5.2   Parallel HashSieve Algorithms

The HashSieve algorithm iterates through the following steps: (1) It randomly samples a lattice vector $\mathbf{v}$ (or fetches one from the stack S). (2) The algorithm identifies a neighboring candidate vector $\mathbf{w}$ from the hash table to perform reduction on $\mathbf{v}$. (3) Subsequently, the algorithm employs the vector $\mathbf{v}$ after reduction to perform reduction on another vector $\mathbf{w}$ from the hash table (if $\mathbf{w}$ undergoes reduction, it is moved onto the stack); (4) Ultimately, the algorithm includes $\mathbf{v}$ in either the stack or the hash table. The HashSieve uses LSH to construct T buckets, i.e., $H_1, ..., H_T$ tables. Each table only stores pointers corresponding vectors and stores them in arrays. At the same time, only one thread is modifying a bucket.

Inspired by [30], and HashSieve is faster than GuassSieve in low dimensions. [28] proposed a scalable parallel implementation and handles concurrency based on a possible lock-free system. An important point is to combine (2), (3) into one step. The algorithm can directly use $\mathbf{v}$ to reduce $\mathbf{w}$ without waiting for $\mathbf{v}$ to be reduced. The latch mechanism proposed in [28] is to set a value for each vector, "0" means that no thread is using this vector, and "1" means that a thread is reading or writing to the vector. If contention occurs, subsequent threads will ignore this vector. This can lead to ignored reductions, but this problem is not serious. Because the probability of contention is low, and these ignored reductions may not be suitable for reduction. In addition, due to the increase in dimensions, the number of buckets will increase exponentially, which may cause only the first iteration to be executed without thread blocking. Therefore, at this time, multiple threads need to be added to operate on the same bucket. When the lock-free mechanism is enabled, the performance of this algorithm is approximately 2.2 times higher compared to GuassSieve, and when the lock-free mechanism is disabled, the performance is approximately 2.5 times higher. The main disadvantage of this algorithm is the large amount of memory used.

5.3   Parallel LDSieve Algorithms

The design of the buckets in LDSieve is straightforward: it randomly selects a direction in space and includes a vector in the bucket if its normalized inner product with vector $\mathbf{c}$ exceeds a certain constant α, expressed as α: $\frac{\langle \mathbf{v}, \mathbf{c} \rangle}{||\mathbf{v}|| \cdot ||\mathbf{c}||} > \alpha$. Unlike employing T randomly chosen vectors $\mathbf{c_1}, \ldots, \mathbf{c_T}$, [8] introduces a different approach by selecting a random subcode $S \subset R^{n/m}$. The code words $C = \{\mathbf{c_1}, \ldots, \mathbf{c_T}\}$ are defined as the product code $C = S \times S \times \ldots \times S = S^m$, where a code word $\mathbf{c_i}$ is constructed by concatenating m code words $\mathbf{c_{i_1}}, \ldots, \mathbf{c_{i_m}}$ from S.

LDSieve first randomly generates a subcode S of dimension n/m and size $t^{1/m}$, defining a concatenated product code $C = S^m$. When processing the vector $\mathbf{v}$, LDSieve calculates the partial dot products among its m sub-vectors with the complete code C, storing the results in lists $L_1, \ldots, L_m$. The decoding algorithm efficiently sorts each list based on the inner product size. Subsequently, within a tree-like structure resembling a depth-first search, it identifies all buckets associated with a vector $\mathbf{c}$ in C that satisfy the condition of the normalized inner product between $\mathbf{c}$ and $\mathbf{v}$ being greater than α.

Also, inspired by [28–30] proposed a scalable parallel iteration of LDSieve featuring a probabilistic lock-free mechanism. The proposed variant relaxed certain properties of the algorithm to accommodate parallelism, demonstrating that LDSieve scales effectively on shared memory systems and utilizes significantly less memory than HashSieve, all while achieving comparable or even reduced execution time on random lattices. And LDSieve ($2^{0.09n}$) stores fewer pointers than GuassSieve ($2^{0.129n}$). Although contention increases as the dimension grows, the number of buckets in LDSieve is sufficiently large that the actual lock is rarely used. Subsequently, [29] made some optimizations to the computation and storage of vectors. They also employed software-based prefetching and manual insertion of prefetch instructions to hide memory request latency. In the end, this algorithm was tested to be 50 times faster than the original implementation of LDSieve.

With the development of GPUs, Tensor cores make their appearance. [14] proposed two parallel bucketing methods, BGJ-like bucketing and BDGL-like bucketing, and a parallel reducing method using Tensor cores.

For BGJ-like bucketing(triple_gpu), the approach initially selects m bucket centers, preferably uniformly distributed on the sphere within the database. Subsequently, each vector $\mathbf{v}$ belonging to the set L in the database is linked with its respective bucket B_{k_v}, where $k_v = \underset{1 \leq k' \leq m}{argmax} |\langle \mathbf{b}'_\mathbf{k}/\|\mathbf{b}'_\mathbf{k}\|, \mathbf{v}\rangle|$. [14] slightly relaxed this condition, allowing vectors to be associated with up to M buckets. In every iteration, fresh bucket centers are selected, normalized, and stored on each individual GPU. Subsequently, the method streams through the GPU for processing the database $\mathbf{v_1}, ..., \mathbf{v_N}$. The distribution of bucket centers is allocated among 16 threads, and each thread is assigned to store only the optimal bucket encountered for each vector. Finally, the best M buckets are returned.

The bucketing approach for BDGL, resembling the asymptotically optimal method outlined in [8], shares similarities with bgj1 and relies on spherical caps. Similar to BGJ-like bucketing, it consistently returns the M best bucket centers for each vector. In the GPU implementation, each warp concurrently handles multiple vectors, with $\mathbf{v_i}$ stored in thread i $(mod32) \in \{1, .., 32\}$. Depending on the dimensionality, the approach employs the Hadamard transform on the initial 32 or 64 coefficients, extracting 32 or 64 inner products simultaneously. The method can randomly permute the coordinates of the vector before the Hadamard transform to acquire additional bucket centers. To minimize branches and reduce overhead, each thread stores only the best-encountered bucket for every vector it handles. Consequently, for each vector, thread i, where i ranges from 1 to 32, stores the optimal buckets among i, $i + 32$, $i + 64$, and so forth. Out of these 32 results, the approach chooses the best M buckets.

In the reducing phase, calculating the inner product between all pairs is essentially equivalent to calculating one-half of the matrix product $Y_t Y$. Thus, utilizing Tensor cores significantly accelerates matrix computation, enhancing reduction speed. The obtained results facilitate filtering close pairs using two inequalities from [14]. Reported in [14], new records were achieved in the SVP Darmstadt Challenge, reaching dimension 180, surpassing the previous record of 155.

6   SVP Challenge

The SVP Challenge is a competition aimed at addressing the Shortest Vector Problem (SVP), as the hardness of SVP is closely related to the security of lattice-based cryptography. As a result, it has captured the attention of numerous researchers.

Based on dimensions and random seeds set by participants, the SVP Challenge generates lattice basis matrices of Goldstein-Mayer type. Participants are tasked with finding an approximate exact solution to the shortest vector problem for the provided lattice basis. To enter the SVP Challenge Hall of Fame, participants need to provide an approximate exact solution shorter than the current record vector in the same dimension or a solution in a higher-dimensional lattice.

The current top 10 participants in the SVP Challenge Hall of Fame are presented in Table 4. The highest SVP Challenge record dimension is 186, accomplished on Jul. 25, 2023. This record was achieved using a sieving algorithm with GPU acceleration, taking 50.3 d. Analyzing the top 10 records shows a consistent trend: all algorithms utilized sieving. Furthermore, nearly all implementations use GPUs for acceleration.

The top 10 records have primarily been broken by several distinct teams. The first three records were achieved by L. Wang and B. Wang, who employed the pnj-BKZ during the preprocessing. Meanwhile, Sun and Chang, along with Ducas et al., used the algorithm introduced by Ducas et al. [14] in their work. The ninth-ranked record was established by L. Wang and Y. Wang [40].

In summary, sieving, G6K, pnj-BKZ, and GPU-based parallel acceleration are the key methods for solving high-dimensional SVP in the current landscape.

Table 4. Top 10 of the SVP Challenge Hall of Fame, data from [32].

Pos	Dim	Norm	Contestant	Alg	Date	GPU	Time
1	186	3484	L. Wang, B. Wang	Sieving	20230725	✓	50.3d
2	184	3494	L. Wang, B. Wang	Sieving	20230703	✓	43.2d
3	182	3444	L. Wang, B. Wang	Sieving	20230528	✓	35.6d
4	181	3544	Y. Sun, S. Chang	Sieving	20230717	✓	31d
5	180	3499	L. Wang, B. Wang	Sieving	20230526	✓	28.1d
6	180	3509	L. Ducas, M. Stevens, W. van Woerden	Sieving	20210208	✓	51.6d
7	178	3447	L. Ducas, M. Stevens, W. van Woerden	Sieving	20210208	✓	22.8d
8	176	3487	L. Ducas, M. Stevens, W. van Woerden	Sieving	20201013	✓	12.5d
9	170	3378	L. Wang, Y. Wang	Sieving	20221003	-	-
10	170	3438	L. Ducas, M. Stevens, W. van Woerden	Sieving	20200512	✓	8d

Acknowlegments. This work is funded by Major Project of Scientific and Technological R&D of Hubei Agricultural Scientific and Technological Innovation Center [grant number 2020–620–000–002–03], Natural Science Foundation of Hubei Province [grant number 2023AFB394], Knowledge Innovation Program of Wuhan-Shuguang Project [grant number 2022010801020283].

References

1. Ajtai, M.: Generating hard instances of lattice problems. In: Proceedings of the twenty-eighth annual ACM symposium on Theory of computing, pp. 99–108 (1996)
2. Ajtai, M., Kumar, R., Sivakumar, D.: A sieve algorithm for the shortest lattice vector problem. In: Proceedings of the thirty-third annual ACM symposium on Theory of computing, pp. 601–610 (2001)
3. Albrecht, M.R., Ducas, L., Herold, G., Kirshanova, E., Postlethwaite, E.W., Stevens, M.: The General Sieve Kernel and New Records in Lattice Reduction. In: Ishai, Y., Rijmen, V. (eds.) EUROCRYPT 2019. LNCS, vol. 11477, pp. 717–746. Springer, Cham (2019). https://doi.org/10.1007/978-3-030-17656-3_25
4. Andrzejczak, M., Gaj, K.: A Multiplatform Parallel Approach for Lattice Sieving Algorithms. In: Qiu, M. (ed.) ICA3PP 2020. LNCS, vol. 12452, pp. 661–680. Springer, Cham (2020). https://doi.org/10.1007/978-3-030-60245-1_45
5. Aono, Y., Nguyen, P.Q., Seito, T., Shikata, J.: Lower Bounds on Lattice Enumeration with Extreme Pruning. In: Shacham, H., Boldyreva, A. (eds.) CRYPTO 2018. LNCS, vol. 10992, pp. 608–637. Springer, Cham (2018). https://doi.org/10.1007/978-3-319-96881-0_21
6. Aono, Y., Wang, Y., Hayashi, T., Takagi, T.: Improved Progressive BKZ Algorithms and Their Precise Cost Estimation by Sharp Simulator. In: Fischlin, M., Coron, J.-S. (eds.) EUROCRYPT 2016. LNCS, vol. 9665, pp. 789–819. Springer, Heidelberg (2016). https://doi.org/10.1007/978-3-662-49890-3_30
7. Bai, S., Laarhoven, T., Stehlé, D.: Tuple lattice sieving. LMS J. Comput. Math. **19**(A), 146–162 (2016)
8. Becker, A., Ducas, L., Gama, N., Laarhoven, T.: New directions in nearest neighbor searching with applications to lattice sieving. In: Proceedings of the twenty-seventh annual ACM-SIAM symposium on Discrete algorithms, pp. 10–24. SIAM (2016)
9. Burger, M., Bischof, C., Krämer, J.: A new parallelization for p3enum and parallelized generation of optimized pruning functions. In: 2019 International Conference on High Performance Computing and Simulation (HPCS), pp. 931–939. IEEE (2019)
10. Burger, M., Bischof, C., Krämer, J.: p3Enum: A new parameterizable and shared-memory parallelized shortest vector problem solver. In: Rodrigues, J.M.F., Cardoso, P.J.S., Monteiro, J., Lam, R., Krzhizhanovskaya, V.V., Lees, M.H., Dongarra, J.J., Sloot, P.M.A. (eds.) ICCS 2019. LNCS, vol. 11540, pp. 535–542. Springer, Cham (2019). https://doi.org/10.1007/978-3-030-22750-0_48
11. Correia, F., Mariano, A., Proenca, A., Bischof, C., Agrell, E.: Parallel improved schnorr-euchner enumeration SE++ for the CVP and SVP. In: 2016 24th Euromicro International Conference on Parallel, Distributed, and Network-Based Processing (PDP), pp. 596–603. IEEE (2016)
12. Dagdelen, Ö., Schneider, M.: Parallel Enumeration of Shortest Lattice Vectors. In: D'Ambra, P., Guarracino, M., Talia, D. (eds.) Euro-Par 2010. LNCS, vol. 6272, pp. 211–222. Springer, Heidelberg (2010). https://doi.org/10.1007/978-3-642-15291-7_21
13. Ducas, L.: Shortest vector from lattice sieving: a few dimensions for free. In: Nielsen, J.B., Rijmen, V. (eds.) EUROCRYPT 2018. LNCS, vol. 10820, pp. 125–145. Springer, Cham (2018). https://doi.org/10.1007/978-3-319-78381-9_5
14. Ducas, L., Stevens, M., van Woerden, W.: Advanced Lattice Sieving on GPUs, with Tensor Cores. In: Canteaut, A., Standaert, F.-X. (eds.) EUROCRYPT 2021. LNCS, vol. 12697, pp. 249–279. Springer, Cham (2021). https://doi.org/10.1007/978-3-030-77886-6_9

15. Esseissah, M.S., Bhery, A., Daoud, S.S., Bahig, H.M.: Three strategies for improving shortest vector enumeration using GPUS. Sci. Program. **2021**, 1–13 (2021)
16. Fincke, U., Pohst, M.: Improved methods for calculating vectors of short length in a lattice, including a complexity analysis. Math. Comput. **44**(170), 463–471 (1985)
17. Gama, N., Nguyen, P.Q., Regev, O.: Lattice enumeration using extreme pruning. In: Gilbert, H. (ed.) EUROCRYPT 2010. LNCS, vol. 6110, pp. 257–278. Springer, Heidelberg (2010). https://doi.org/10.1007/978-3-642-13190-5_13
18. Ghasemmehdi, A., Agrell, E.: Faster recursions in sphere decoding. IEEE Trans. Inf. Theory **57**(6), 3530–3536 (2011)
19. Hermans, J., Schneider, M., Buchmann, J., Vercauteren, F., Preneel, B.: Parallel shortest lattice vector enumeration on graphics cards. In: Bernstein, D.J., Lange, T. (eds.) AFRICACRYPT 2010. LNCS, vol. 6055, pp. 52–68. Springer, Heidelberg (2010). https://doi.org/10.1007/978-3-642-12678-9_4
20. Ishiguro, T., Kiyomoto, S., Miyake, Y., Takagi, T.: Parallel gauss sieve algorithm: solving the SVP challenge over a 128-dimensional ideal lattice. In: Krawczyk, H. (ed.) PKC 2014. LNCS, vol. 8383, pp. 411–428. Springer, Heidelberg (2014). https://doi.org/10.1007/978-3-642-54631-0_24
21. Kannan, R.: Improved algorithms for integer programming and related lattice problems. In: Proceedings of the fifteenth annual ACM symposium on Theory of computing, pp. 193–206 (1983)
22. Kuo, P.-C., Schneider, M., Dagdelen, Ö., Reichelt, J., Buchmann, J., Cheng, C.-M., Yang, B.-Y.: Extreme enumeration on GPU and in clouds. In: Preneel, B., Takagi, T. (eds.) Extreme enumeration on GPU and in clouds. LNCS, vol. 6917, pp. 176–191. Springer, Heidelberg (2011). https://doi.org/10.1007/978-3-642-23951-9_12
23. Laarhoven, T.: Sieving for shortest vectors in lattices using angular locality-sensitive hashing. In: Gennaro, R., Robshaw, M. (eds.) CRYPTO 2015. LNCS, vol. 9215, pp. 3–22. Springer, Heidelberg (2015). https://doi.org/10.1007/978-3-662-47989-6_1
24. Laarhoven, T., Mariano, A.: Progressive lattice sieving. In: Lange, T., Steinwandt, R. (eds.) Progressive lattice sieving. LNCS, vol. 10786, pp. 292–311. Springer, Cham (2018). https://doi.org/10.1007/978-3-319-79063-3_14
25. Laarhoven, T., Mosca, M., van de Pol, J.: Solving the shortest vector problem in lattices faster using quantum search. In: Gaborit, P. (ed.) PQCrypto 2013. LNCS, vol. 7932, pp. 83–101. Springer, Heidelberg (2013). https://doi.org/10.1007/978-3-642-38616-9_6
26. Lenstra, A.K., Lenstra, H.W., Lovász, L.: Factoring polynomials with rational coefficients. Mathematische annalen **261**(ARTICLE), 515–534 (1982)
27. Loyer, J., Chailloux, A.: Classical and quantum 3 and 4-sieves to solve SVP with low memory. Cryptology ePrint Archive (2023)
28. Mariano, A., Bischof, C., Laarhoven, T.: Parallel (probable) lock-free hash sieve: a practical sieving algorithm for the SVP. In: 2015 44th International Conference on Parallel Processing, pp. 590–599. IEEE (2015)
29. Mariano, A., Laarhoven, T., Bischof, C.: A parallel variant of lDSieve for the SVP on lattices. In: 2017 25th Euromicro International Conference on Parallel, Distributed and Network-based Processing (PDP), pp. 23–30. IEEE (2017)
30. Mariano, A., Timnat, S., Bischof, C.: Lock-free gausssieve for linear speedups in parallel high performance SVP calculation. In: 2014 IEEE 26th International Symposium on Computer Architecture and High Performance Computing, pp. 278–285. IEEE (2014)

31. Micciancio, D., Voulgaris, P.: Faster exponential time algorithms for the shortest vector problem. In: Proceedings of the twenty-first annual ACM-SIAM symposium on Discrete Algorithms, pp. 1468–1480. SIAM (2010)
32. Michael, S., Nicolas, G.: SVP Challenge (2010). https://www.latticechallenge.org/SVP-challenge/
33. Milde, B., Schneider, M.: A parallel implementation of gausssieve for the shortest vector problem in lattices. In: Malyshkin, V. (ed.) PaCT 2011. LNCS, vol. 6873, pp. 452–458. Springer, Heidelberg (2011). https://doi.org/10.1007/978-3-642-23178-0_40
34. Mukhopadhyay, P.: Faster provable sieving algorithms for the shortest vector problem and the closest vector problem on lattices in l_p norm. Algorithms **14**(12), 362 (2021)
35. Nguyen, P.Q., Vidick, T.: Sieve algorithms for the shortest vector problem are practical. J. Math. Cryptology **2**(2), 181–207 (2008)
36. Pohst, M.: On the computation of lattice vectors of minimal length, successive minima and reduced bases with applications. ACM Sigsam Bulletin **15**(1), 37–44 (1981)
37. Schnorr, C.P., Euchner, M.: Lattice basis reduction: improved practical algorithms and solving subset sum problems. Math. Program. **66**, 181–199 (1994)
38. Schnorr, C.P., Hörner, H.H.: Attacking the Chor-Rivest cryptosystem by improved lattice reduction. In: Guillou, L.C., Quisquater, J.-J. (eds.) EUROCRYPT 1995. LNCS, vol. 921, pp. 1–12. Springer, Heidelberg (1995). https://doi.org/10.1007/3-540-49264-X_1
39. Tateiwa, N., et al.: Massive parallelization for finding shortest lattice vectors based on ubiquity generator framework. In: SC20: International Conference for High Performance Computing, Networking, Storage and Analysis, pp. 1–15. IEEE (2020)
40. Wang, L., Wang, Y., Wang, B.: A trade-off SVP-solving strategy based on a sharper PNJ-BKZ simulator. In: Proceedings of the 2023 ACM Asia Conference on Computer and Communications Security, pp. 664–677 (2023)
41. Wang, X., Liu, M., Tian, C., Bi, J.: Improved nguyen-vidick heuristic sieve algorithm for shortest vector problem. In: Proceedings of the 6th ACM Symposium on Information, Computer and Communications Security, pp. 1–9 (2011)
42. Xia, W., Wang, L., Gu, D., Wang, B., et al.: Improved progressive BKZ with lattice sieving. Cryptology ePrint Archive (2022)
43. Schnorr, C.P., Hörner, H.H.: Attacking the Chor-Rivest Cryptosystem by Improved Lattice Reduction. In: Guillou, L.C., Quisquater, J.-J. (eds.) EUROCRYPT 1995. LNCS, vol. 921, pp. 1–12. Springer, Heidelberg (1995). https://doi.org/10.1007/3-540-49264-X_1
44. Yasuda, M.: A survey of solving SVP algorithms and recent strategies for solving the SVP challenge. In: Takagi, T., Wakayama, M., Tanaka, K., Kunihiro, N., Kimoto, K., Ikematsu, Y. (eds.) A survey of solving SVP algorithms and recent strategies for solving the svp challenge. MI, vol. 33, pp. 189–207. Springer, Singapore (2021). https://doi.org/10.1007/978-981-15-5191-8_15
45. Zhang, F., Pan, Y., Hu, G.: A Three-level sieve algorithm for the shortest vector problem. In: Lange, T., Lauter, K., Lisoněk, P. (eds.) A three-level sieve algorithm for the shortest vector problem. LNCS, vol. 8282, pp. 29–47. Springer, Heidelberg (2014). https://doi.org/10.1007/978-3-662-43414-7_2

Blockchain and Applications

PSWS: A Private Support-Weighted Sum Protocol for Blockchain-Based E-Voting Systems

Chenyu Deng[✉]

School of Computer Science, Hubei University of Technology, Wuhan, China
csmwzhang@gmail.com

Abstract. Nowadays, e-voting systems are receiving a lot of attention. Voters can cast their votes for the candidates they support on the e-voting system. However, how to reflect the voters' support level for the candidates in the e-voting system in a fair and privacy-preserving way is still a problem that needs to be solved. This article proposes a private support-weighted sum (PSWS) protocol, which is a fair, and privacy-preserving weighted sum protocol. The PSWS protocol privately calculates the weighted sum of each voter's support degree for a candidate. After the protocol's execution, others on the system can only access the weighted sum, without any additional information. The PSWS protocol has two novel characteristics. Firstly, the voting terminal or polling station provides encryption services for ballots immediately after each ballot is cast. All voting information is expressed in ciphertext throughout the weighting and counting processes, until the final result of the weighted vote is passed to the decryption server in ciphertext. This design avoids the disclosure of voter privacy and ballot information, and the ciphertext format also prevents malicious users from cheating or tampering with voter or ballot information during the counting process. Security analysis is conducted to validate security properties. Secondly, our protocol not only achieves the function of privacy-preserving support-weighted voting but also is relatively lightweight and efficient. Finally, the efficiency analysis results of the experiments in terms of calculation show that the protocol meets the requirements applicable to real-world applications. In summary, PSWS can be harmoniously applied to candidate election in blockchain systems.

Keywords: Privacy protection · Homomorphic encryption · Weighted sum protocol · Blockchain

---

This work is supported in part by the Major Research Plan of Hubei Provience under Gran No. 2023BAA027, and the Key Research and Development Program of Hubei Province under Grant 2021BEA163.

J. Chen and Z. Xia (Eds.), BlockTEA 2023, LNICST 577, pp. 79–93, 2024.
https://doi.org/10.1007/978-3-031-60037-1_5

1   Introduction

In today's increasingly digital world, the demand for secure and transparent e-voting systems increasingly critical [1]. Electronic voting machine hacking, vote rigging, electoral fraud, and capturing the polling booth are some of the major issues in current election system [15]. Ensuring the integrity of elections is fundamental to upholding the principles of democracy and maintaining public trust in the electoral process [16]. Traditional voting methods have faced numerous challenges, including concerns about voter fraud, ballot tampering, and the logistical complexities of conducting elections. To address these challenges, emerging technologies such as blockchain offer a promising avenue for revolutionizing the way we conduct and secure elections. Blockchain is totally transparent, secure and immutable technique because it uses concepts like encryption, decryption, hash function, consensus and Merkle tree [9].

Let us consider the following application scenario within electronic voting systems. We have a voter who wishes to cast votes for multiple candidates in an election. The voter supports several candidates to varying degrees but is unwilling to disclose his private information. In such a scenario, the problem to address is how to enable voting with varying degrees of support for multiple candidates without revealing individual voters' preferences or compromising privacy. Therefore, this paper proposes a privacy-preserving weighted sum protocol, private support-weighted sum(PSWS). If the critical problem underlying an application scenario can be translated to the privacy-preserving aggregation problem of a series of integer arrays (each array can be viewed as the support degree received by each candidate), then the PSWS protocol applies to this scenario.

1.1   Contributions

The main contributions of our work can be summarized as follows:

- To address the challenge of facilitating a voter's ability to cast votes for multiple candidates within a blockchain-based election voting system, while accurately reflecting the level of support for each candidate, we propose an innovative solution: the PSWS protocol.
- Expanding upon our proposed solution, we introduce a privacy-preserving support-weighted voting algorithm that prioritizes privacy protection while accurately computing the overall support degree for each candidate. This algorithm incorporates advanced techniques, such as homomorphic encryption, to enable the computation of candidate support degree without compromising individual voters' privacy. This approach guarantees the accurate determination of total support degree for each candidate while respecting the confidentiality of voters' choices.
- We undertake an extensive security analysis to validate the effectiveness of the PSWS protocol in preserving both the privacy of user data and computation results. This rigorous analysis serves to establish the algorithm's resilience against potential attacks and its capacity to offer a substantial level of security in safeguarding user data privacy while upholding the confidentiality of computed results.

1.2   Plan of This Paper

The remaining structure of this paper is organized as follows: In Sect. 2, we provide some preliminary insights. In Sect. 3, we present the system model, threat model, security requirements, and design goals. Section 4 showcases the specific details of the system design, followed by security proofs in Sect. 5. Section 6 covers the experimental part, including performance evaluation. In Sect. 7, we delve into related work, and in Sect. 8, we conclude the entire paper while providing prospects for future work.

2   Preliminaries

In this section, we reviewed the relevant knowledge of BGN [4] homomorphic encryption and blockchain that were utilized throughout the entire work. As shown in Table 1, we outline the notations used in this paper.

Table 1. Table captions should be placed above the tables.

Notation	Remarks
τ	Security parameter
e	Bilinear map
Can	Set of candidates
n	Number of voters
u	Number of candidates
$a_{i,j}$	The jth vote for the ith candidate
t_i	The ith candidate supported by the voters
$m_{i,j}$	The support degree paid by the jth voter of the ith candidate
$[[\frac{1}{n}]]$	Ciphertext of $\frac{1}{n}$
C_{t_i}	Ciphertext of t_i
$[[\frac{1}{n^2}]]$	Ciphertext of $\frac{1}{n^2}$
$C_{m_{i,j}}$	Ciphertext of $m_{i,j}$
V_{m_i}	The set of ciphertexts of the support degrees for candidate t_i
T_i	Sum of support degrees for the ith candidate
C_{T_i}	Ciphertext of T_i
Var_i	Variance of support degrees for the ith candidate
C_{Var_i}	Ciphertext of Var_i
$score(t_i)$	The sum of the weighted support degree of the candidate t_i

2.1 BGN Homomorphic Encryption Scheme

The BGN homomorphic cryptosystem is composed of the following:

- **Gen**(τ): Input the security parameter τ and running **Gen**(τ) yields the tuple (q_1, q_2, G, G_1, e), where G and G_1 are groups of order $N = q_1 * q_2$, $e : G * G \rightarrow G_1$ is a bilinear map, and two random generators $k, U \leftarrow G$. Setting $h = U^{q_2}$ results in h being a random generator of the q_1-order subgroup of group G. Set the public key $PK = (N, G, G_1, e, h, k)$ and the private key $SK = q_1$.
- **Encryption** (PK, m): Give a plaintext space $\{0, 1, 2, ..., T\}(T < q_2)$, and given a random selection of $r_N \xleftarrow{R} \{0, 1, ..., N - 1\}$, along with the input plaintext message m and the public key PK, the output ciphertext $C = k^m h^r \in G$ is generated.
- **Decryption** (SK, C): Input the ciphertext C and the private key SK, you can calculate $C^{q_1} = (k^m h^r)^{q_1} = (k^{q_1})^m$. By utilizing Pollard's lambda algorithm [6] to solve the discrete logarithm modulo k^{q_1}, you can recover the plaintext message m.

  BGN encryption scheme has the following homomorphic property:

  Homomorphic Addition: $C = c_1 c_2 h^r = k^{m_1 + m_2} h^{r_1 + r_2 + r}$.

  Homomorphic Multiplication: $C = e(C_1, C_2) h_1^r = k_1^{m_1 m_2} h_1^r$.

2.2 Blockchain

Blockchain is the underlying technology of a number of digital cryptocurrencies [10]. It is a decentralized, distributed ledger technology that offers transparency, security, and trust in a digital world plagued by vulnerabilities and mistrust [5].

At the heart of blockchain lies decentralization. Unlike traditional centralized systems, where a single authority controls data and transactions, blockchain relies on a network of nodes [2]. This decentralized structure ensures no single entity has unilateral control, making it resistant to censorship and tampering [13].

Every transaction on the blockchain is recorded in a public ledger, accessible to all participants in the network [11]. This transparency fosters trust, as anyone can verify the integrity of transactions independently [12].

In conclusion, blockchain technology represents a groundbreaking innovation with the potential to transform the way data is stored, shared, and transacted in various industries [19]. Its decentralized, transparent, and secure nature has led to a burgeoning body of research and practical implementations, shaping the future of digital transactions and trust mechanisms [18].

2.3 Support-Weighted Voting Algorithm

In order to reflect the level of voter support for a candidate, we introduce a support-weighted voting algorithm. The support-weighted voting algorithm is described in detail below. Suppose that the set of u candidates is denoted as $Can = \{t_1, t_2, ..., t_u\}$. For each candidate $t_i \in Can$, a total of n voters cast their votes for it, and the corresponding support degree's are denoted as the set $M_{t,i} = \{m_{i,1}, m_{i,2}, ..., m_{i,n}\}$, where $m_{i,j}$ denotes the support degree paid by the jth voter of the ith candidate for that vote. The support score for each candidate was calculated according to the following formula.

$$score(t_i) = \frac{\sum_{j=1}^{n} m_{i,j}}{1 + Var} \quad (1)$$

$$Var_i = \frac{1}{n} \sum_{j=1}^{n} (\widehat{m_i} - m_{i,j})^2 \quad (2)$$

$$\widehat{m_i} = \frac{1}{n} \sum_{j=1}^{n} m_{i,j} \quad (3)$$

Equation 1 calculates the candidate's support score using the variance as a bias term. Equation 2 calculates the variance of support degree. Equation 3 calculates the everage value of support degree. This support voting algorithm with variance as a bias term is used mainly to solve the problem of choosing the best result among multiple candidates who have received votes with the same support degree. In the case of multiple candidates receiving the same total vote score, using variance as a bias term tends to select results with similar support votes from different voters. That is, we believe that the result of a vote in which the voters are more in agreement is more likely to be the accurate result. To facilitate subsequent calculations, we change the variance to the following form:

$$Var_i = \frac{\sum_{j=1}^{n} m_{i,j}^2}{n} - \frac{(\sum_{j=1}^{n} m_{i,j})^2}{n^2} \quad (4)$$

3   Models and Design Goals

In this section, We focus on system model, threat model and design goals.

3.1   System Model

There exist five types of entities in the system: Blockchain, Voter, Encryption Server, Decryption Server and Canditate, which is shown in Fig. 1

- **Blockchain:** The blockchain is responsible for storing voting and ballot information, and the content of each ballot, as well as the corresponding support degree for each candidate, is recorded on the blockchain.
- **Canditates:** Candidates create ballots and submitting them to the blockchain in order to allow voters to cast their votes.
- **Voters:** Each voter can vote for any of the candidates on the ballot, and the ballot includes the candidate the voter supports and the support degree for that candidate. Then, the voter needs to encrypt and upload the voting information to the encryption server using the public key distributed by the decryption server.

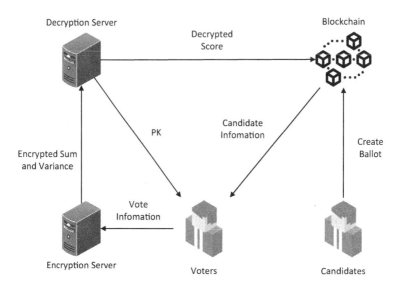

Fig. 1. System model.

– **Encryption Server**: The encryption server is responsible for aggregating the support degree uploaded by the voter over the ciphertext through homomorphic encryption. The sum and variance of the aggregated support degree is passed to the decryption server for further processing.
– **Decryption Server**: The decryption server decrypts the sum and variance of each candidate's support degree and then uses a support-weighted voting algorithm to calculate the total support degree for each candidate. After that, it is uploaded to the blockchain.

3.2   Threat Model and Security Requirements

In our threat model, Voters are assumed to be semi-honest (i.e., they follow the protocol while harboring a curiosity regarding each other's multi-dimensional data). Similarly, the encryption server is also considered semi-honest (i.e., it adheres to the protocol but exhibits curiosity towards the data shared by Voters). Furthermore, external attackers have the ability to eavesdrop on communications in order to obtain transmitted reports. In this context, we require that the decryption server cannot collude with any other entities.

In this paper, we regard the voting data of voters as their private information. We further emphasize the requirement that the aggregated support degree for each candidate, obtained through weighted aggregation, remain private until the voting results are made public. The following security requirements should be met:

– The encryption server and external attackers are unable to obtain the voting data of voters.
– External adversaries are unable to access detailed voting results before they are publicly disclosed.

3.3   Design Goal

Under the aforementioned system model, threat model, and security require-ments, our design objective is to develop a privacy-preserving support-weighted voting algorithm. Specifically, we aim to achieve the following two desirable goals.

- The proposed privacy-preserving support-weighted voting algorithm should meet the defined security requirements. Failure to do so may result in poten-tial violations of voters' privacy, leading to a lack of willingness among voters to provide their true support degree data during the voting process.
- Through the privacy-preserving support-weighted voting algorithm, the blockchain is able to obtain the accurate weighted aggregation result of sup-port for each candidate.

4   The Proposed Methodology

To achieve the PSWS protocol, it is necessary to provide an algorithm that enables the computing party to calculate the weighted voting results without decrypting the voting data of the voters.

4.1   A Privacy-Preserving Support-Weighted Voting Algorithm

To fulfill the security requirements mentioned, we incorporated a homomorphic encryption scheme (such as BGN encryption) into the support-weighted voting algorithm, thus forming the proposed privacy-preserving support-weighted vot-ing algorithm. The framework of the privacy-preserving support-weighted voting algorithm is illustrated in Fig. 2. Note that the figure shows the flow of one round of voting in the algorithm, i.e., it shows n voters voting for one candidate t_i(the ith candidate). If there are n voters who want to vote on u candidates, then it is only necessary to carry out the process in the diagram for the u round. This framework consists of five stages. In the first stage, the blockchain sends information about the candidates publicly to all voters. In the second stage, the decryption server sends the public key PK to all voters. In the third stage, the voters encrypt their votes using the public key PK and then package them and send them to the encryption server. In the fourth stage, the encryption server calculates the sum and the variance of the candidate's support degree on the ciphertext and sends them to the decryption server. In the fifth stage, the decryption server decrypts the sum as well as the variance of the candidate's support degree and calculates the candidate's total support *score* using Eq. 1 and sends it to the blockchain for public disclosure.

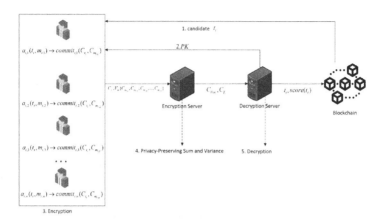

Fig. 2. The framework of the proposed voting algorithm

Algorithm 1. Encryption

Input: $a_{i,j} = (t_i, m_{i,j}), PK = (N, G, G_1, e, h, k)$
Output: $commit_{i,j} = (C_{t_i}, C_{m_{i,j}})$
1: Random $r \leftarrow \{0, 1, ..., N-1\}$
2: Compute $C_{t_i} = k^{C_{t_i}} h^r, C_{m_{i,j}} = k^{m_{i,j}} h^r$
3: Return $commit_{i,j} = (C_{t_i}, C_{m_{i,j}})$;

In Algorithm 1, $a_{i,j}$ denotes the jth vote for the ith candidate, where t_i denotes the ith candidate supported by the voters, and $m_{i,j}$ denotes the support degree paid by the jth voter of the ith candidate for that vote. The value of $m_{i,j}$ is in the range of $[0,10]$. 10 indicates full support for the candidate and 0 indicates no support at all for the candidate. If the voter wants to indicate no support at all for the candidate, then the support degree in the vote is entered as 0. PK is the public key of the BGN homomorphic encryption algorithm generated by the decryption server. This algorithm utilizes this public key to encrypt the voter's vote $a_{i,j}$. The algorithm's output $commit_{i,j}$ is an encrypted representation of a vote.

In Algorithm 2, C_{t_i} represents the ciphertext of a candidate obtained after encryption using the BGN algorithm in Algorithm 1. V_{m_i} denotes the collection of ciphertexts representing the support degrees received by this candidate. C_{T_i} represents the sum of all support degrees in V_{m_i} in ciphertext form. It can be further represented using the BGN homomorphic encryption algorithm as follows:

$$C_{T_i} = k^{\sum_{j=1}^{n} m_{i,j}} \times h^{\sum_{j=1}^{n} r_j + r} \tag{5}$$

where k and h are both generators used in the BGN homomorphic encryption algorithm. The $C_{i,j}$ in step 3 of Algorithm 2 represents the square of each support degree in V_{m_i} in ciphertext form. It can be further represented using the BGN

Algorithm 2. Privacy-Preserving Sum and Variance

Input: $C_{t_i}, V_{m_i} = \{C_{m_{i,1}}, C_{m_{i,2}}, C_{m_{i,3}}, ..., C_{m_{i,n}}\}$

Output: $Sum_i = (C_{t_i}, C_{Var_i}, C_{T_i})$

1: Compute $C_{T_i} = \prod\limits_{j=1}^{n} C_{m_{i,j}}$

2: **for** $j = 1 \rightarrow n$ **do**

3:     Compute $C_{i,j} = e(C_{m_{i,j}}, C_{m_{i,j}})$

4: **end for**

5: Compute $C_{Var_i} = \dfrac{[\![\frac{1}{n}]\!] \prod\limits_{j=1}^{n} C_{i,j}}{[\![\frac{1}{n^2}]\!] e(C_T, C_T)}$

6: Return $Sum_i = (C_{t_i}, C_{Var_i}, C_{T_i})$

---

homomorphic encryption algorithm as follows:

$$C_{i,j} = k_1^{m_{i,j}^2} h_1^{\tilde{r}}. \tag{6}$$

The C_{Var_i} in step 4 of Algorithm 2 represents the variance of all support degrees in V_{m_i} in ciphertext form. It can be further represented using the BGN homomorphic encryption algorithm as follows:

$$C_{Var_i} = \frac{[\![\frac{1}{n}]\!] k^{\sum\limits_{j=1}^{n} m^2_{i,j}} \times h^{\sum\limits_{j=1}^{n} r_j + r}}{([\![\frac{1}{n^2}]\!] e(k^{\sum\limits_{j=1}^{n} m_{i,j}} \times h^{\sum\limits_{j=1}^{n} r_j + r}, k^{\sum\limits_{j=1}^{n} m_{i,j}} \times h^{\sum\limits_{j=1}^{n} r_j + r})}. \tag{7}$$

This algorithm is designed to calculate the variance of support degree C_{Var_i} and the sum of support degree C_{T_i} on the ciphertext for a candidate. Importantly, the entire process is performed on ciphertext, significantly enhancing the algorithm's security.

Algorithm 3. Decryption

Input: $Sum_i = (C_{t_i}, C_{Var_i}, C_{T_i}), SK = q_1$

Output: $s_i = (t_i, score(t_i))$

1: Compute $(C_{t_i})^{q_1} = (k^{q_1})^{t_i}$

2: Compute $(C_{Var_i})^{q_1} = (k^{q_1})^{Var_i}$

3: Compute $(C_{T_i})^{q_1} = (k^{q_1})^{T_i}$

4: $t_i, Var_i, T_i \leftarrow$ Using Pollard's lambda algorithm[6] to compute the discrete logarithm with base k^{q_1}.

5: Compute $score(t_i) = \dfrac{T_i}{1 + Var_i}$

6: Return $s_i = (t_i, score(t_i))$

Algorithm 3 is designed to decrypt the variance of support degree C_{Var_i} and the sum of support degreeC_{T_i} on the ciphertext for a candidate. Then, the variance was used as a bias term in order to calculate the sum *score* of the weighted support degree of the candidate t_i.

Algorithm 4. Privacy-preserving Support-weighted Voting Algorithm

---

Input: $V_a = \{\{a_{1,1}, a_{1,2}, ..., a_{1,n}\}, \{a_{2,1}, a_{2,2}, ..., a_{2,n}\}, ..., \{a_{u,1}, a_{u,2}, ..., a_{u,n}\}\}, PK = (w, G, G_1, e, h, k), SK = q_1$

Output: $S = \{(t_1, score(t_1)), (t_2, score(t_2)), ..., (t_u, score(t_u))\}$

1: $S = \{\}, V_{m_i} = \{\}, Tmp = \{\}$
2: **for** $a_{i,j} = (t_i, m_{i,j}) \in V_a, i \in [1, u], j \in [1, n]$ **do**
3:     $commit_{i,j} = Encryption(a_{i,j}, PK)$
4:     $Append(V_{m_i}, C_{m_{i,j}})$
5: **end for**
6: $Tmp = \{(C_{t_i}, V_{m_i})|(C_{t_1}, V_{m_1}), (C_{t_2}, V_{m_2}), ..., (C_{t_u}, V_{m_u})\}$
7: **for** $(C_{t_i}, V_{m_i}) \in Tmp$ **do**
8:     $Sum_i = Privacy\text{-}Preserving\ Sum\ and\ Variance(C_{t_i}, V_{m_i})$
9:     $s_i = Decryption(Sum_i, SK)$
10:     $Append(S, s_i)$
11: **end for**
12: Return S

The complete privacy-preserving support-weighted voting algorithm is shown in Algorithm 4. V_a represents the set of votes of the voters, each round of voting has only one identified candidate, the number of rounds is the number of candidates u. This means that the system needs to allow all voters to vote in u rounds, and the candidate for each round is uniquely determined.

5   Security Property Analysis

According to our security requirements, we discuss how the proposed protocol PSWS ensures the privacy protection of the voters during the weighted aggregation of their votes.

Theorem 1. *The proposed protocol PSWS enables the privacy protection of voting data for the voters.*

Proof. As the voters first invoke the decryption server to generate a public key PK and encrypt their voting data with it, they subsequently send the encrypted ciphertext data to the encryption server for homomorphic weighted aggregation operations. Other nodes on the blockchain cannot access the voting data of the voters because they are unaware of the decryption server's private key SK. After aggregating the ciphertexts of the voters' support degrees using the Privacy-Preserving Support-Weighted Voting Algorithm, the encryption server sends the aggregated ciphertext to the decryption server. The voting information is encrypted under the BGN cryptosystem. The BGN cryptosystem based on compound-order bilinear groups is proven to be semantically safe against selected plaintext attacks. In the computation on encryption server, many homomorphic operations are also encrypted under the BGN cryptosystem, so it is impossible to deduce any meaningful content.

Consequently, the proposed Privacy-Preserving Support-Weighted Voting Algorithm accomplishes the privacy protection of the voters' voting data.

6   Experiments

In this chapter, we empirically discussed the feasibility and practicality of the PSWS protocol through experimental analysis. In our simulation experiments, we employed BGN homomorphic encryption to cryptographically secure the support degrees obtained for each candidate by the encryption server. In other words, the ciphertext utilized in the Privacy-Preserving Support-Weighted Voting Algorithm was of the BGN cryptographic form. We assumed the existence of u candidates, with each candidate receiving n randomly generated votes of support. $Time_1$ and $Time_2$ represented the number of encryption and decryption operations in the BGN encryption scheme, respectively. We used Z_n to denote the computation time for weighted aggregation operations.

We randomly generated voting data for 5 candidates, with each candidate receiving 5 votes, each containing a support degree, for the purpose of conducting simulation experiments. We conducted the simulations on a computer equipped with a 3.40GHz 13th Gen Intel(R) Core(TM) i7-13700KF processor, 32GB of RAM, and a 64-bit operating system. For the BGN encryption scheme, we adopted a security parameter length of $256bits$. The experimental results indicated that $Time_1$ was 595ms, $Time_2$ was 69ms, and the Privacy-Preserving Support-Weighted Voting Algorithm could be completed within 1823ms. Among these, the running time for the aggregation operation, denoted as Z_n, was approximately 1210 ms. The support degrees obtained by the 5 candidates in this experiment are shown in Table 2. The voting results, as depicted in Fig. 3, indicate that Candidate 5 receives the highest total support degree, making Candidate 5 the winner of this vote.

Table 2. Case

candidate	support degree				
	supportdegree1	supportdegree2	supportdegree3	supportdegree4	supportdegree5
candidate1	7	9	2	4	9
candidate2	5	4	7	10	5
candidate3	8	3	7	8	1
candidate4	4	6	1	3	5
candidate5	5	7	7	4	8

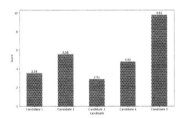

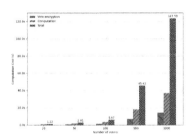

Fig. 3. Score for each candidate.

Fig. 4. Time overhead analysis of our protocol deployment ($t = 256bit$).

Figure 4 reflects the approximate linear correlation between the number of voters and the time overhead of communication. It is attributed to the number of voters will increase the number of operations and ciphertext.

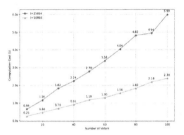

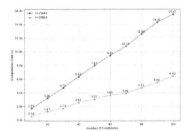

Fig. 5. Time overhead with different number of voters in different security parameter.

Fig. 6. Time overhead with different number of candidates in different security parameter.

The analysis of the time overhead is shown in Fig. 5 and Fig. 6. In Fig. 5, the number of candidates $n_c = 1$. In Fig. 6, the number of voters $n_v = 1$. Through the comparison of Fig. 5 and Fig. 6, it can be found that, since the impact of the security parameters on the encryption operation (i.e., complex encryption and increased ciphertext length), the time overhead is positively correlated with the security parameters.

7   Related Work

In [14], G. Revathy *et al.* proposed an electronic voting scheme for face recognition based on deep learning technology.Daria Golnarian *et al.* [7]proposed a novel trustless e-voting scheme based on blockchain technology that minimizes the intervention of authorities in the process, and voters can vote remotely using their own mobile devices. Hemlata Kohad *et al.* [8]proposed the multiobjective genetic algorithm-based creation of a side-chain to enhance the scalability and performance of the blockchain-based e-voting system. Blockchain employs cryptographic techniques to secure data and transactions [17]. Information is stored in blocks, which are cryptographically linked in a chain. Once a block is added to the chain, altering its contents would require altering all subsequent blocks, making it nearly impossible to manipulate past data [20]. Based on the requirements analysis, Md Jobair Hossain Faruk *et al.* [3] proposed a biometric-enabled and hyperledger fabric-based voting framework to automate identity verification that will ensure transparency and security of electronic voting.

In contrast to the aforementioned work, our framework places a strong emphasis on conducting elections effectively and fairly while ensuring privacy protection throughout the entire process. The highlight of our work is the proposal of a voting algorithm framework that preserve the privacy of voting data while using weighted support degrees as the measurement criterion, and we implement this framework using homomorphic encryption methods.

8   Conclusion

This article proposes a privacy-preserving support-weighted sum protocol (PSWS). This protocol allows voters to cast their votes for the candidates they support. Voters can express their support for multiple candidates by assigning different support degrees to each candidate. The protocol employs the Privacy-Preserving Support-Weighted Voting Algorithm to calculate the votes, resulting in a fair and effective election outcome. Importantly, it ensures full privacy protection throughout the entire process, safeguarding the confidentiality of voters' voting data and preliminary results until the election results are allowed to be made public. To implement the privacy-preserving support-weighted voting algorithm, we utilize BGN homomorphic encryption technology. We prove that our proposed privacy-preserving support-weighted voting algorithm meets the security requirements within the threat model. Allowing supporters to vote for multiple candidates not only provides them with more choices but also enhances the fairness and usability of the voting system. However, it also results in increased computational overhead for privacy-preserving support-weighted voting algorithms. Future research can explore methods to enhance the computational efficiency of the current system, reduce computational costs, and mitigate the mentioned weaknesses. Furthermore, we plan to explore weighted voting with multidimensional data in more complex models, such as scenarios involving multiple evaluation criteria.

References

1. Chatterjee, U., Ray, S., Adhikari, S., Khan, M.K., Dasgupta, M.: Efficient and secure e-voting scheme using elliptic curve cryptography. Secur. Priv. **6**(3), e283 (2023)
2. Decker, C., Wattenhofer, R.: Information propagation in the bitcoin network. In: IEEE P2P 2013 Proceedings, pp. 1–10. IEEE (2013)
3. Faruk, M.J.H., et al.: Development of blockchain-based e-voting system: Requirements, design and security perspective. In: 2022 IEEE International Conference on Trust, Security and Privacy in Computing and Communications (TrustCom), pp. 959–967. IEEE (2022)
4. Freeman, D.: Homomorphic encryption and the BGN cryptosystem (2011)
5. Gadekallu, T.R., et al.: Blockchain for the metaverse: a review. arXiv preprint.arXiv:2203.09738 (2022)
6. Gallant, R., Lambert, R., Vanstone, S.: Improving the parallelized pollard lambda search on anomalous binary curves. Math. Comput. **69**(232), 1699–1705 (2000)
7. Golnarian, D., Saedi, K., Bahrak, B.: A decentralized and trustless e-voting system based on blockchain technology. In: 2022 27th International Computer Conference, Computer Society of Iran (CSICC), pp. 1–7. IEEE (2022)
8. Kohad, H., Kumar, S., Ambhaikar, A.: Scalability of blockchain based e-voting system using multiobjective genetic algorithm with sharding. In: 2022 IEEE Delhi Section Conference (DELCON), pp. 1–4. IEEE (2022)
9. Kumar, D., Dwivedi, R.K.: Designing a secure e voting system using blockchain with efficient smart contract and consensus mechanism. In: International Conference on Advanced Network Technologies and Intelligent Computing, pp. 452–469. Springer (2022)
10. Monrat, A.A., Schelén, O., Andersson, K.: A survey of blockchain from the perspectives of applications, challenges, and opportunities. IEEE Access **7**, 117134–117151 (2019)
11. Niranjanamurthy, M., Nithya, B., Jagannatha, S.: Analysis of blockchain technology: pros, cons and swot. Clust. Comput. **22**, 14743–14757 (2019)
12. Omran, Y., Henke, M., Heines, R., Hofmann, E.: Blockchain-driven supply chain finance: Towards a conceptual framework from a buyer perspective. Int. Purchas Supply Educ. Res. Assoc. **2017**, 15–15 (2017)
13. Rauchs, M., et al.: Distributed ledger technology systems: a conceptual framework. Available at SSRN 3230013 (2018)
14. Revathy, G., Raj, K.B., Kumar, A., Adibatti, S., Dahiya, P., Latha, T.: Investigation of e-voting system using face recognition using convolutional neural network (cnn). Theoret. Comput. Sci. **925**, 61–67 (2022)
15. Vijayakumar, K., Sriramasivam, T., Vigneshkumar, G., Senthil Kumar, G., Angel, T., Snehalatha, N.: Secure e-voting system using blockchain based on solidity technology. In: AIP Conference Proceedings. vol. 2516. AIP Publishing (2022)
16. Wahab, Y., : A framework for blockchain based e-voting system for Iraq. Int. J. Interact Mobile Technol. **16**(10) (2022)
17. Zhai, S., Yang, Y., Li, J., Qiu, C., Zhao, J.: Research on the application of cryptography on the blockchain. In: J. Phys.: Conference Series. vol. 1168, p. 032077. IOP Publishing (2019)
18. Zhang, M., Yang, M., Shen, G.: Ssbas-fa: A secure sealed-bid e-auction scheme with fair arbitration based on time-released blockchain. J. Syst. Architect. **129**, 102619 (2022) 10.1016/j.sysarc.2022.102619, https://www.sciencedirect.com/science/article/pii/S1383762122001503

19. Zhang, M., Yang, M., Shen, G., Xia, Z., Wang, Y.: A verifiable and privacy-preserving cloud mining pool selection scheme in blockchain of things. Inf. Sci. **623**, 293–310 (2023) 10.1016/j.ins.2022.11.169, https://www.sciencedirect.com/science/article/pii/S0020025522015225
20. Zhang, R., Xue, R., Liu, L.: Security and privacy on blockchain. ACM Comput. Surv. (CSUR) **52**(3), 1–34 (2019)

Cross-Chain Trusted Information Match Scheme with Privacy-Preserving and Auditability

Zheng Chen[iD], Zejun Lu[iD], and Jiageng Chen[(✉)][iD]

Central China Normal University, Wuhan 430079, Hubei, China
`jiageng.chen@ccnu.edu.cn`

Abstract. Recently, numerous challenges concerning personal identity security along with other security issues regarding privacy have been introduced. While the blockchain offers a solution for recording personal information and facilitating matching needs, existing researches have indicated the absence of privacy and traceability. In this paper, we introduce an information matching scheme rooted in cross-chain technology. This approach employs a public blockchain for information matching and a consortium blockchain for transmitting private data. Cross-chain consensus protocols are leveraged to accomplish privacy protection and validation for cross-chain information transmission. The scheme serves the purpose of achieving information matching between two parties within the public blockchain framework. It enhances users' identity and information privacy, while also bolstering the auditability of cross-chain information.

Keywords: Blockchain · Cross-chain · Privacy preserving

1 Introduction

With the advent of informatization and the new era, the evolving nature of information matching stands to significantly enhance global communication quality. Ensuring the privacy of individuals' information during the information exchange process remains a pivotal concern in this information matching paradigm. The following issues deserve attention: (1) Security of Identity and Information: The inevitable disclosure of both parties' identities lies at the heart of information matching, potentially compromising the privacy of their respective information. (2) Recognition of Information from Different Parties: Insufficient data sharing among distinct parties often leads to a failure in recognizing exchanged information, hindering effective communication. (3) Auditability of Published Information: The immutability of blockchains can lead to situations where malicious information is published by senders, negatively impacting the integrity of information matching. Addressing these challenges will be crucial in establishing a robust and effective information matching model that upholds privacy and communication quality.

© ICST Institute for Computer Sciences, Social Informatics and Telecommunications Engineering 2024
Published by Springer Nature Switzerland AG 2024. All Rights Reserved
J. Chen and Z. Xia (Eds.), BlockTEA 2023, LNICST 577, pp. 94–114, 2024.
https://doi.org/10.1007/978-3-031-60037-1_6

Blockchain is a tamper-resistant digital ledger implemented in a distributed environment and typically without a central authority. At a fundamental level, it enables a community of users to upload transactions to a shared ledger at a relatively low cost. Under the normal operation of the blockchain network, once published, no transaction can be altered, rendering the blockchain immutable. Blockchain users can transmit information and transfer their rights to another user. The blockchain publicly documents this transfer process through transactions, allowing all participants in the blockchain network to independently verify the transaction's validity.

The rise and development of blockchain technology, represented by Bitcoin [1] and PrCash [2], introduces a new scenario unlike traditional platforms. After years of in-depth development, the blockchain has witnessed a multitude of coexisting variations, each possessing distinct characteristics suitable for various application scenarios. However, a regular decentralized blockchain does not meet certain usage scenarios in information matching where all information should be stored privately, and its efficiency is also relatively low. With the rapid growth of the blockchain industry and the emergence of numerous public chains, private chains, and consortium chains, a challenge arises: how to facilitate communication and even value exchange between different blockchains. Due to the segregation of blockchains and the significant heterogeneity between different blockchains, the transmission of information and data communication among existing blockchains faces unprecedented challenges. This situation also gives rise to numerous difficulties in facilitating communication between different blockchains.

The concept of cross-chain was initially proposed by the Tendermint team in 2014. In a narrower sense, cross-chain involves asset interoperability between relatively independent blockchain ledgers. In a broader sense, it encompasses data and asset interoperability between independent blockchain ledgers. Traditionally, interoperability in the information field refers to 'the ability to exchange and use information between different systems or modules.' [3] In terms of communication between blockchains, 'Chain Interoperability' mainly concerns the ability to transfer assets, make payments, or exchange information between two blockchains. This can be accomplished by introducing third parties without altering the native chain. A single blockchain network is relatively closed and lacks active external interaction. Cross-chain technology strives to build trust bridges between chains, dismantling the notion that a blockchain is akin to an isolated island. Cross-chain technology within the blockchain domain serves as a critical method for achieving interconnection, enhancing interoperability, and bolstering scalability between blockchains. In cross-chain industry applications, such as Celer cBridge [4], MultiChain [5], Synapse Bridge [6], and Umbria Narni [7], cross-chain technologies can support the cross-chain transmission of blockchain information across various blockchain networks. Additionally, these technologies can also find applications in certain industrial scenarios. Moreover, they possess the capability to accommodate an expanding array of blockchain types,

1.1 Our Contribution

In comparison to existing research solutions, existing information matching solutions primarily operate on a single blockchain, such as public blockchains, which may not be suitable for certain scenarios. This can result in low throughput, reduced efficiency (in terms of query and record times), and a lack of privacy protection for personal information.

In this article, we present an information storage and matching scheme that leverages cross-chain technology, incorporating public blockchains, consortium education blockchains, and cross-chain consensus protocols.

(1) In our solution, personal information storage is conducted on a consortium blockchain, while information matching is performed on a public blockchain. This approach not only avoids frequent interactions with the public blockchain during information queries but also ensures efficient recording and retrieval of personal information on the consortium blockchain.
(2) The consortium blockchain facilitates the secure storage of users' private information. This information is verified through a Zero-knowledge proof provided by certified consortium blockchain members, and it can relatively reduce blockchain storage costs. Notably, an auditor is in place to unveil potentially maliciously stored information and identify the responsible member by decrypting transactions on the consortium blockchain.
(3) The public blockchain executes the matching of information between two public users. It serves to offer information matching functionality while extending security properties to both users involved in a transaction.

By employing this comprehensive approach, our scheme seeks to enhance information management, matching, and security through the synergy of cross-chain technology, consortium education blockchains, and public blockchains.

1.2 Organization

The remainder of our work is organized as follows. First, we introduce the related works in Sect. 2. The preliminaries are presented in Sect. 3. The scheme is presented in Sect. 4. Then, we describe the details of our protocol and security properties in Sect. 5. Finally, in Sect. 6, we conclude this work.

2 Related Works

In cross-chain scenarios, [8] proposes a cross-chain application called Collafab. It uses the concept of 'collect-sign' [9] to create a cross-chain consensus based on PBFT [10] on the private blockchain consensus. Additionally, it supports interoperability between both public and private users on two blockchains. Current schemes and protocols have certain weaknesses and shortcomings. There are issues with data communication and sharing among different parties, including challenges such as unverifiable records and the risk of uncontrolled information transfer and matching.

In cross-chain application scenarios, A. Garoffolo et al. [11] proposed Zendoo, a cross-chain transfer protocol that enables decoupled and decentralized sidechain creation and communication. It extends the functionality of a blockchain system like Bitcoin as the main blockchain by utilizing smart contracts, while sidechains support different types of consensus and transactions. However, Zendoo's basic functions rely on the main blockchain, thus lacking decentralization, and the model of the sidechain provides low security and lacks user information privacy. It only adapts to a certain type of blockchain, and the relationship between the main blockchain and sidechains leads to low scalability. Yin et al. [12] proposed an open distributed secure cross-chain notary platform based on MPC (BOOL network). This platform utilizes a Ring-Verifiable random function to securely and verifiably deliver secret keys to new members in the blockchain. This approach helps in concealing the identity of the current owner while enhancing platform compatibility and user-friendliness, all the while maintaining the openness of the public blockchain. It claims that this platform supports different types of blockchains and compatibility perfectly. However, BOOL network is unable to support auditability. He et al. [13] designed a cross-chain-based computing scheme (TSQC) that deploys smart contracts in each blockchain to manage inner chain computation. This strategy reduces cross-chain information exchange, resulting in improved system efficiency. Moreover, the scheme employs the CoSi protocol and multisigncryption algorithms for safeguarding cross-chain data privacy. This scheme supports at least three blockchains and is capable of scalability; however, there is a risk of information leakage except for interchain interoperation. Yi et al. [14] proposed a cross-chain-based premium competition scheme with privacy preservation (CCUBI). This scheme employs a bridge contract and a third-party trust agent to facilitate cross-chain information transfer. User clients generate proofs of data that ensure privacy and computability in the cross-chain scenario. The framework of this scheme can be implemented on two types of blockchains. However, for a malicious node which executes aggregation verification, there is a risk of user information leakage. Kuongho Chen et al. [15] introduced a trusted reputation management scheme (TRM) that achieves cross-chain communication through a relay chain. The relay chain handles cross-chain transactions and consensus, managing the sending and receiving of requests from all nodes in the blockchain system. However, this scheme incurs a high time cost due to the deployment of smart contracts and cross-chain consensus in the blockchain it employs. Although relay chain-based blockchain perfectly supports multiple blockchains, it also lacks decentralization and scalability, and the property of anonymity is unable to provide information security.

Finally, Table 1 compares different cross-chain schemes in terms of several properties with our scheme. Here is the following explanation of the table: Decentralization denotes that information can be transferred without a central party. Info-privacy denotes that no one knows the information value except the sender,

Table 1. Comparison among cross-chain schemes.

Scheme \ Feature	Decentralization	Info-privacy	Scalability	Compatibility	Auditability
Zendoo	◑	○	◒	◑	○
BOOL network	●	●	●	●	○
TSQC	●	◑	◒	◑	○
CCUBI	●	◑	◒	◑	●
TRM	◑	○	◑	●	○
This work	●	●	◑	◑	●

●: the scheme perfectly confirms to the following feature or supports all types of blockchains.
◑: the scheme partly confirms to a feature or supports a certain type of blockchain.
◒: the scheme strictly supports a certain blockchain.
○: the scheme dose not support the feature.

receiver, and the auditor. Moreover, scalability is vital for a cross-chain scheme that supports scenarios involving more than just two blockchains. Compatibility means the scheme is capable of working with different types of blockchains. Auditability means that transactions on the blockchain can be audited by a certain trusted node.

3   Preliminaries

3.1   Pedersen Commitment [16]

A sender can commit a value to create a commitment with randomness to the receiver. Both the sender and the receiver can compute the value by evaluating the commitment. The sender can later disclose the actual value by opening the commitment using the message and the randomness [17]. In elliptic curve cryptography, we denote the Pedersen commitment as $C \leftarrow \text{Cmt}(v, r)$, where v is the message, and r is the randomness. Given randomness r and the commitment C, one can recover the value $v \leftarrow \text{Open}(C, r)$.

3.2   Boneh-Boyen Signature [18]

Let $\mathbb{G}_1, \mathbb{G}_2$ be bilinear groups where $|\mathbb{G}_1| = |\mathbb{G}_2|$ for prime p. Let public parameter g_1 be a generator of $\mathbb{G}_1$ and g_2 be a generator of $\mathbb{G}_2$. To sign a message $m \in Z_p$:

- $(pk, sk) \leftarrow \text{BBKeyGen}(1^\lambda)$: input a security parameter, the algorithm outputs a secret/public key pair.
- $\sigma \leftarrow \text{BBSign}(sk, m)$: input a secret key sk and for input a message $m \in \{0, 1\}^*$, it outputs a signature σ of message m.
- $1/0 \leftarrow \text{BBVer}(pk, m, \sigma)$: input a public key pk, a message m and a BBS signature σ, the algorithm outputs 1 if σ is valid and outputs 0 otherwise.

3.3   Aggregate Signatures

Boneh et al. introduced BLS aggregate signatures [19] and introduced the concept of aggregate signatures [20] as follows:

Let $\mathbb{G}_1, \mathbb{G}_2$ be bilinear groups where $|\mathbb{G}_1| = |\mathbb{G}_2|$ for prime p, and g_1 be a generator of $\mathbb{G}_1$ and g_2 be a generator of $\mathbb{G}_2$. H is a full-domain collision resistance hash function $H : \{0,1\}^* \to \mathbb{G}_1$.

- KeyGen (1^λ) : input a parameter 1^λ, the algorithm outputs a secret/public key pair (sk, pk).
- $\sigma \leftarrow$ Sign (sk, M) to sign a message $M \in \{0,1\}^*$ and input a secret key sk, it outputs a signature denotes as $\sigma \in \mathbb{G}_1$.
- $1/0 \leftarrow$ Ver (pk, M, σ) : the algorithm output 1 if σ is a valid signature and outputs 0 otherwise.
- $\sigma \leftarrow$ AggSig (σ_0, σ_1) Input two BLS signatures (σ_0, σ_1), the algorithm outputs an aggregate signature σ.
- $1/0 \leftarrow$ AggSigVer $(\mathbf{L}, \sigma)$. Input n tuples $\mathbf{L} : \{(pk_i, m_i)\}^n$ and a signature σ, the algorithm outputs 1 if the aggregate signature σ is valid to all messages and outputs 0 otherwise.

3.4   Zero-Knowledge Range Proof

The concept of range proof was first introduced by Boneh et al. in 2008 [21]. Then Bunz et al. [22] introduced a zero-knowledge proof scheme. To prove a secret v in range $[0, u^l]$, the secret v can be decomposed to $\varepsilon = \sum_{0 \le j \le l}(\varepsilon_j u^j)$. We define the zero-knowledge range proof scheme as (ZKProve, ZKVer).

- $\pi \leftarrow$ ZKProve (v, k) where v is the input secret, and k is corresponding randomnesses, it output a zeroknowledge proof $\pi : (C_s, \hat{C}_s, \pi_s)$.
- $1/0 \leftarrow$ ZKVer (π). Input a zero-knowledge range proof $\pi : (C_s, \hat{C}_s, \pi_s)$, it outputs 1 if π is a valid proof for the value v.

4   System Model

4.1   Scheme Construction

We formalize our scheme by extending the functionalities of Hyperledger Fabric [23] and Aggregate Cash system [24]. We define several types of nodes in our scheme below:

- Consortium member (CM): CM can generate response messages in response to requests from the trust agent and subsequently submit transactions to the consortium blockchain. The transaction is created for cross-chain consensus by utilizing their individual secret keys.
- Trust agent: The trust agent has the capacity to engage with both the consortium blockchain and public users. It can transfer information requests directly to the consortium blockchain, access transactions stored within the consortium blockchain, and send cross-chain information to users.

- Certificate Authority (CA): CA holds the authority to issue certificates to members of the consortium. This issuance ensures that only authorized consortium members possess the privilege to submit transactions to the consortium blockchain.
- Auditor: The auditor possesses the capability to decrypt any transaction within the consortium blockchain, thus retrieving the details of the transaction.
- Public user: Users are able to initiate information requests to the consortium blockchain through the trust agent. They can broadcast a desired target value and anticipate its reception, pre-match transactions to the designated receiver, and submit matching transactions to the public blockchain.

4.2   Security Threats

We assume that the system parameters are generated. We consider the privacy properties of the attacker model below:

The nodes of CA are honest but curious. They honestly run the protocol while recording all inputs and outputs. The auditor strictly obeys the protocol.

Attackers can act as malicious nodes in both CMs and public users.

Attackers know all parameters and public keys of public users and CMs.

Our scheme focuses on several security properties as follows:

Sybil Attacks. As mentioned above, attackers can manipulate, control, or generate malicious nodes among public users. These nodes are generated to attack the system. However, the consensus algorithm of the public blockchain (varies depending on the type of blockchain) can resist Sybil attacks.

Privacy. The identities of the two parties involved in public matching cannot be known by public blockchain users except themselves, nor can they be known by other consortium members except the one responding to the request. User private information for public matching cannot be leaked during transmission in the public blockchain; it is only known by the consortium member that generates it, the two matching parties, and the authority.

Unforgeability. Attackers cannot generate legitimate cross-chain information, consensus, or public chain matching information by forging any secret information, signatures, or proofs in an attempt to damage the system.

The details of our scheme's security are shown in Sect. 5.3.

5   Our Protocol

Our basic construction is shown in Fig. 1. In this section, we describe the process of a transaction in our scheme in Sect. 4.1, followed by the detailed construction in Sect. 4.2. The security of our scheme is discussed in Sect. 5.3.

5.1 Workflow

- For a user A, they broadcast a target value t on the public channel, expecting other users' private values to match it (they only receive values greater than the target value).
- User B initiates a query request through a trusted agent to a designated consortium member of the consortium blockchain. Their identity is indicated by the public key, and the identity of the consortium member is indicated by its public key.
- The trusted agent verifies user query requests via SPV and then posts them to the consortium blockchain. After the designated consortium members receive the request, they generate the corresponding information as a response: a public blockchain transaction and a secret key encrypted by user B's public key.
- All responses from consortium members must undergo cross-chain consensus signature, and the trusted agent only receives transactions signed by the majority of consortium members, which it then submits to the public blockchain. Similarly, user B only receives the ciphertext under the same conditions and decrypts it to obtain the key corresponding to the value in the transaction.
- If the value corresponding to the secret key received by user B is greater than the target t, a matching request message is initiated to user A, which is encrypted with A's public key and sent to A.
- User A decrypts the matching request message, verifies its correctness, and checks if m is greater than or equal to t. Then, they generate a matching transaction while setting their own secret key corresponding to the received value m, and submit the transaction to the public blockchain.
- The Auditor can reveal transactions from the consortium blockchain, recover details in the transaction, and obtain user information from the transaction.

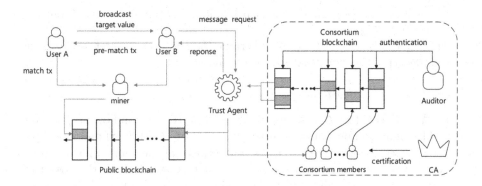

Fig. 1. Basic construction of our scheme

5.2 System Construction

System Initialization

- $(pp, R) \leftarrow$ Setup (1^λ). Input a security parameter 1^λ, and the algorithm outputs the system parameters pp, generates a tuple $(p, g, g_0, \hat{g}, \mathbb{G}_1, \mathbb{G}_2, \mathbb{G}_T, e)$ where $(g, g_0) \in G_1, \hat{g} \in g_2$, the bilinear map $e : G_1 \times G_2 \rightarrow \mathbb{G}_T$ and a range R for zero-knowledge proofs.
- A collision resistance hash function $H : \{0, 1\}^* \rightarrow \mathbb{G}_1$.

Consortium blockchain algorithms
  Following consortium algorithms are inspired by Pachain [25]:

- $(apk, ask, capk, cask) \leftarrow$ AuKeyGen (). The algorithm generates the auditor's and the CA's public/secret key pairs.
- $(mpk, msk) \leftarrow$ CMKeyGen (1^λ). This algorithm outputs consortium members' public/secret key pairs (mpk, msk).
- The auditor generates a bitmap $\mathbb{B}$ that indicates which members signed the message for an aggregate signature. Each member who signed the message should add their bit in the proper location by $\mathbb{B}_i \leftarrow$ NewBit $(\mathbb{B})$.
- $(\Pi : (C_s, \pi_s)) \leftarrow$ EncProof (m). Input a message m, the algorithm generates a ciphertext C_s and a zero-knowledge proof π_s.
- $1 \leftarrow$ EncVer $(\Pi : (C_s, \pi_s))$.Input a ciphertext C_s and a proof π_s, the algorithm outputs 1 if Π is a valid zero-knowledge proof.
- $m \leftarrow$ EncDec $(\Pi : (C_s, \pi_s), ask)$. Input the auditor secret key ask and a ciphertext C_s and a proof π_s, the auditor can decrypt C_s to get the value m where the value is in the range of $[0, R]$.

Public CCA Encryption Scheme

- $(upk, usk) \leftarrow$ PKeyGen (pp). Input a security parameter pp, it outputs a user public/secret key pair (upk, usk).
- $C_m \leftarrow$ Enc (M, upk).Input a message M and a user public key upk, it outputs a ciphertext C_m.
- $M \leftarrow$ Dec (C_m, ask). "Input a ciphertext C_m and a user secret key usk; the user decrypts the ciphertext C_m to get the message.

Public Blockchain Algorithms

- $(tx, \hat{k}) \leftarrow$ CreateTx $((C, v, k), \hat{v})$. Input the tuple (C, v, k) and the output score $\hat{v}$. The algorithm outputs a transaction $tx = (s, C, \hat{C}, \Pi, E, \sigma)$ and an output key $\hat{k}$, where $\hat{C}$ denotes the output commitment of the transaction.
- $1/0 \leftarrow$ VerTx (tx). Input a transaction $tx = (s, \mathbf{C}, \hat{\mathbf{C}}, \Pi, E, \sigma)$, the algorithm outputs 1 if tx is a valid transaction and outputs 0 otherwise.
- $(tx) \leftarrow$ AggTx (tx_0, tx_1). Input two transactions (tx_0, tx_1), the algorithm puts them together and returns an aggregate transaction tx.

Message Transfer Algorithms

- PostPB (): This algorithm allows a public user (including the trust agent) to submit a transaction to the public blockchain.
- PostCB (): This algorithm allows a consortium member (including the trust agent) to submit a message to the consortium blockchain.
- SendtoU (): This algorithm allows a public user (including the trust agent) to send a message to other users in a public channel.
- SendtoA (): This algorithm allows a public user to send a message to the trust agent in a public channel.

Cross-Chain Consensus Construction   In this section, we describe the construction as follows. The details of the construction are shown in Fig. 2.

The process begins with a consortium member CM_0 initiating cross-chain consensus on a public transaction. This member starts by creating a public transaction denoted as tx and generates the corresponding key k for the message m. Subsequently, it triggers a cross-chain message and posts it onto the consortium blockchain.

For any consortium member, denoted as CM_i where $i \in [1, N]$, that reads a cross-chain message from the consortium blockchain, submitted by another user, the member validates the message and signs it using an aggregate signature mechanism. This generates a cross-chain consensus facilitated by the secret key msk_i.

The trust agent is responsible for extracting messages from the consortium blockchain. It selectively accepts messages that satisfy the criteria of cross-chain consensus, subsequently submitting the public transaction tx to the public blockchain.

Cross-Chain Information Response to Users   In this section, we describe the construction of the cross-chain response to the user as follows. The details of the construction are shown in Fig. 3.

To achieve cross-chain privacy for user information, there is a distinction from cross-chain public information (such as public blockchain transactions). In this scheme, the user's requested consortium blockchain information is encrypted by a designated consortium member and subsequently dispatched to the user.

User B initiates a message request directed at a specific consortium member denoted as CM_l, identified by the public key mpk_l, and transmits it to the trust agent. The agent verifies this request and then submits it onto the consortium blockchain.

The consortium member CM^*_0 gets the message from the consortium blockchain and creates a cross-chain response for user B, which is subsequently submitted onto the consortium blockchain. The encrypted message creation process follows the same cross-chain consensus procedure as that of the tx.

The trust agent reviews the messages present on the consortium blockchain and only accepts messages destined for users that meet the conditions of cross-

IniReq $(upk_B, \hat{mpk}_0, upk_a)$:
- $req \leftarrow Enc ((upk_B, \hat{mpk}_0), upk_a)$.
- SendtoA (req) .

PostReq (req, usk_a):
- $(upk_B, \hat{mpk}_0) \leftarrow Dnc (req, usk_a)$.
- PostCB $(upk_B, \hat{mpk}_0)$.

CreateCss (tx, m, msk_0):
- $tx \leftarrow CreateTx ((C, v, k), m)$: $C \leftarrow Cmt (m, k)$, $\pi \leftarrow ZKProve (m, k)$.
- $\Pi \leftarrow EncProof (m)$.
- $H_{tx} = H(tx, \Pi)$.
- $\sigma_0 \leftarrow Sign (H_{tx}, msk_0)$: $\sigma_0 = H(tx)^{msk_0}$.
- $\mathbb{B}_0 \leftarrow NewBit ()$.
- PostCB $(css_0 = (tx, \Pi, H_{tx}, \mathbb{B}_0, \sigma_0))$.

AggCss $(css_j = (tx, H_{tx}, \Pi, \mathbb{B}_j, \sigma_j), msk_i)$: $(i > j; j \geq 0)$:
- if (
    $H_{tx} \neq H(tx, \Pi)$ or
    $0 \leftarrow AggSigVer (mpk_j, H_{tx}, \sigma_j)$: (j is indicated by $\mathbb{B}_j$) or
) return $\perp$.
- $\sigma' \leftarrow Sign (msk_i, H_{tx})$.
- $\sigma_i \leftarrow AggSig (\sigma', \sigma_j)$.
- $\mathbb{B}_i \leftarrow NewBit (\mathbb{B}_j)$.
- PostCB $(css_i = ((tx, \Pi, H_{tx}, \mathbb{B}_i, \sigma_i))$.

TxSubmt $(ccs_t = (tx, \Pi, H_{tx}, \mathbb{B}_t, \sigma_t))$:
- if (
    $H_{tx} \neq H(tx, \Pi)$ or
    $0 \leftarrow AggSigVer (mpk_t, H_{tx}, \sigma_t)$: (t is indicated by $\mathbb{B}_t$) or
    $0 \leftarrow EncVer (\Pi)$ or
    $|\mathbb{B}| \leq \frac{2}{3}N$ (N is the total number of CM) or
    $0 \leftarrow VerTx (tx)$.
) return $\perp$.
- PostCB (tx).

- IniReq ( ) is executed by the public user B.
- PostReq ( ) is executed by the trust agent.
- CreateCss ( ) is executed by a consortium member.
- AggCss () can be executed by all consotium members.
- TxSubmt () is executed by the trust agent.

Fig. 2. Construction of public transaction cross-chain consensus

chain consensus. Subsequently, the agent sends these messages to their designated recipients.

Finally, user B receives the message, validates its authenticity, and then decrypts the content to obtain their own information represented by m.

CreateCssU $(tx, m, k, \hat{msk}_0, \hat{mpk}_0)$:

- $\tau_k \leftarrow H(upk, \hat{mpk}_0, k, tx)$.
- $k_u \leftarrow$ Enc (k, tx, upk) and $H_k = H(k_u)$.
- $\sigma_\tau \leftarrow$ BBSign $(\hat{msk}_0, \tau_k)$.
- $\hat{\sigma} \leftarrow$ Sign $(\hat{msk}_0, H_k)$.
- $\mathbb{B}_\tau \leftarrow$ NewBit ().
- PostCB $(c\hat{ss}u_0 : (k_u, H_k, \tau_k, \sigma_\tau, \hat{\sigma}, \mathbb{B}_\tau))$.
- $c\hat{ss}u_i \leftarrow$ AggCss $(c\hat{ss}u_j = (k_u, H_k, \tau_k, \sigma_\tau, \hat{\sigma}_j, \mathbb{B}_{\tau_j})) : (i > j; j \geq 0)$.
- PostCB $(c\hat{ss}u_i)$.

ResVer $(c\hat{ss}u_t = (k_u, H_k, \tau_k, \sigma_\tau, \hat{\sigma}_t, \mathbb{B}_{\tau_t}))$:

- if (

    $H_{tx} \neq H(k_u)$ or

    $0 \leftarrow$ AggSigVer $(mpk_j, H_{tx}, \hat{\sigma}_j) :$ (j is indicated by $\mathbb{B}_j$) or

    $|\mathbb{B}| \leq \frac{2}{3} N$ or

    $0 \leftarrow$ BBSVer $(\hat{mpk}_0, \tau_k, \sigma_k)$.

) return $\perp$.
- SendtoU $(cssu_t)$.

UserRcv $(ccsu_t = (k_u, H_k, \tau_k, \sigma_\tau, \hat{\sigma}_t, \mathbb{B}_{\tau_t}))$:

- if (

    $H_{tx} \neq H(k_u)$ or

    $0 \leftarrow$ AggSigVer $(mpk_t, H_{tx}, \hat{\sigma}_t) :$ (j is indicated by $\mathbb{B}_t$) or

    $|\mathbb{B}| \leq \frac{2}{3} N$ or

    $0 \leftarrow$ BBSVer $(\hat{mpk}_0, \tau_k, \sigma_{\tau_k})$.

) return $\perp$.
- $(k, tx) \leftarrow$ Dec (usk_B, k_u).
- $m \leftarrow$ Open (C, k).

- CreateCssU ( ) is executed by consortium members.
- ResVer ( ) is executed by the trust agent.
- UserRcv ( ) is executed by the public user B.

Fig. 3. Construction of user information cross-chain consensus

Public Information Matching   In this section, we describe the construction of the public information match as follows. The details of the construction are shown in Fig. 4.

For a public user A, they generate a target value t and broadcast it to all public users, expecting to receive a value which is $\geq t$. For a public user B, who intends to send their value m satisfying the condition $m \geq t$ to user A while keeping other values denoted as v^*, initiates a Prematch transaction. They then send the encryption of the prematch transaction to user A. Upon receiving a *pmtx* from user B, user A validates it and generates a match transaction along with a new key, effectively granting user A ownership of the information. Finally, user A submits a match transaction to the public blockchain while retaining the secret key privately.

Announce (t):
- SendtoU (t).

PreMatch (upk_A, C, m, k):
- $(tx_b, k_b) \leftarrow$ CreateTx (C, v, k, 0) .
- $C' =$ Cmt (m, k') where $k' \leftarrow Z_p$ at random.
- $pmtx = (tx_b, m, k')$.
- SendtoU (pm = Enc ((pmtx, C'), upk_A)).

Match (pmtx, usk_A):
- $(pmtx = (tx_b, m, k'), C') \leftarrow$ Dec (pm, usk_A).
- if (
    $0 \leftarrow$ TxVer (tx_b) or
    $0 \leftarrow$ ZKVer (π or
    $C' \neq$ Cmt (m, k') $\wedge$ m $\geq$ t
) return $\perp$.
- $(tx_a, k_a) \leftarrow$ CreateTx $((C', m, k'), m_a)$.
- $tx_m \leftarrow$ AggTx (tx_b, tx_a).
- PostCB (tx_m).

- Announce ( ) is executed by the public user A.
- PreMatch ( ) is executed by the public user B.
- Match ( ) is executed by the public user A.

Fig. 4. Construction of public information matching

Auditability   In this section, we describe the construction of the audit as follows. The details of the construction are shown in Fig. 5.

The auditor can decrypt a transaction on the consortium blockchain to recover the necessary transaction details. Notably, the auditor can particularly uncover the identity of the first member who initially submitted the transaction to the consortium blockchain. The auditor validates the message on the consortium blockchain and decrypts the ciphertext to reveal the information m in the transaction tx.

CssDec (ccs_i = (tx, Π, H_{tx}, $\mathbb{B}_i$, σ_i), ask):
- if (
    $H_{tx} \neq H(tx, \Pi)$ or
    $0 \leftarrow$ AggSigVer (mpk_i, H_{tx}, σ_i) : (i is indicated by $\mathbb{B}_i$) or
    $1 \leftarrow$ EncVer (Π : (C_s, π_s)) or
    $|\mathbb{B}| \leq \frac{2}{3}N$ or
    $0 \leftarrow$ VerTx (tx).
) return $\perp$.
- $m \leftarrow$ EncDec (Π, ask).

- CssDec () is executed by the auditor.

Fig. 5. Construction of auditability

5.3   System Security

In this section, we present security theorems concerning the proposed scheme and illustrate their proof by simplifying them into the security of cryptographic primitives. In our subsequent security analysis, public information (such as transactions and public keys within the public blockchain, and the public keys of consortium members) is considered to be known to potential adversaries.

We will first define the security properties of consortium blockchain algorithms in the following section.

Definition 1. *(Soundness) For any P.P.T. adversary $\mathcal{A}$:*

$$Pr[ask \leftarrow \mathcal{A} : (\Pi^* = (\pi_s^*, C_s^*)) \leftarrow \mathcal{A}(ask),$$
$$1 \leftarrow \text{ZKVer}(\pi_s^*) = 1, m \leftarrow \text{EncDec}(C_s^*, ask)] \text{ is negligible.}$$

Thus our consortium blockchain algorithms ensure soundness.

Definition 2. *(CPA Semantically secure indistinguishability)*
For any P.P.T. CPA adversary $\mathcal{A}$, here we define two experiments, Experiment 0 and Experiment 1. In each experiment $b = \{0, 1\}$, the adversary generates m_0, m_1 and sends to the challenger, the challenger computes $\Pi^* \leftarrow$ EncProof (m_b) where $b \leftarrow \{0, 1\}$, and sends it to $\mathcal{A}$, the adversary outputs $\hat{b} \in \{0, 1\}$.

Let Exp_{A_b} be the adversary outputs 1 in experiment b, then the advantage of $\mathcal{A}$ in algorithm EncProof ():

$$Adv_{\mathcal{A}}^{\text{EncProof}} = |Pr[Exp_{A_0}] - Pr[Exp_{A_1}]| \text{ is negligible}$$

Thus our consortium blockchain algorithms are semantically secure against a CPA adversary.

Next, we will define the security properties of public blockchain algorithms:

Definition 3. *(Security against inflation) The only way a message can be created is through cross-chain consensus transactions. This implies that for any transaction, the total value of the outputs should be equal to the sum of the total value of the inputs plus the new value of the transaction. For any P.P.T. adversary $\mathcal{A}$:*

$$Pr[(tx, v) \leftarrow \mathcal{A} : tx' \leftarrow \mathcal{A}(v'), v' > v \wedge 1 \leftarrow \text{TxVer}(tx')] = \text{neg}(\lambda).$$

Definition 4. *(Security against theft) The property of theft resistance ensures that only the key owner can utilize the message for matching and aggregation purposes. For any P.P.T. adversary $\mathcal{A}$:*

$$Pr[tx \leftarrow \mathcal{A} : k^* \leftarrow \mathcal{A}(tx), (tx', \hat{k}') \leftarrow \text{CreateTx}((C, v, k^*), \hat{v})] = \text{neg}(\lambda).$$

Definition 5. *(Transaction indistinguishable)*

The amounts involved in a transaction are concealed, ensuring that only the sender and receiver are aware of the sum of exchanged. Furthermore, a transaction comprehensively conceals the associations between inputs and outputs, making it impossible to discern which inputs funded which outputs.

For any P.P.T. adversary $\mathcal{A}$, the adversary generates $(v_0, \hat{v}_0), (v_1, \hat{v}_1)$ and sends to the challenger, the challenger computes $tx^* \leftarrow \mathrm{CreateTx}\,(v_b, \hat{v}_b)$ where $b \leftarrow \{0, 1\}$, and sneds it to $\mathcal{A}$, the adversary outputs $\hat{b} \in \{0, 1\}$.

Let Exp_{tx_b} be the adversary outputs 1 in experiment b, then the advantage of $\mathcal{A}$ in algorithm CreateTx ():

$$Adv_{\mathcal{A}}^{\mathrm{TxIND}} = |Pr[Exp_{tx_0}] - Pr[Exp_{tx_1}]| \leq \text{ negligible}.$$

Definition 6. *(Correctness). Our scheme has correctness if several following conditions satisfy:*

- *User requests can be correctly decrypted and verified by the trust agent, and consortium members can generate legitimate transactions.*
- *Cross-chain consensus can be accurately executed. Once the signature count surpasses a specified threshold, the recipient can verify cross-chain information using the designated member public keys indicated by the bitmap $\mathbb{B}$.*
- *Zero-knowledge proofs generated by the consortium chain can be verified by any honest node.*
- *Public users are able to correctly decrypt cross-chain encrypted information sent by consortium members and encrypted matching information from other public users.*

Theorem 1. *Our scheme satisfies correctness if all algorithms are correct.*

Proof. The correctness can be obtained from the workflow of the scheme and can be derived from the correctness of the following cryptographic primitives:

We first focus on the abilities of the adversary $\mathcal{A}$, where $\mathcal{A}$ has the following oracles:

- O_{cu}: The adversary can corrupt any user node to obtain (upk, usk) and make queries to the consortium blockchain via the trust agent.
- The adversary can make two types of queries to the CCA public encryption simulator:
  O_{pke}: The adversary can send a message m to the simulator to obtain the encrypted ciphertext c.
  O_{pkd}: The adversary can send a ciphertext c, which was not received in previous encryption queries, to obtain the decryption of the ciphertext.
- O_{cm}: The adversary can corrupt any single consortium member to obtain (msk, mpk), read information on the consortium blockchain, and initiate a cross-chain consensus message.

– $O_{sig}(m)$: The adversary can make queries to an aggregate signature simulator to request a signature on a message signed by the secret key msk.

User query security means that attackers cannot obtain the identities of users and consortium members from queries received by the trust agent. Now, consider the experiment described below in Fig. 6. For a P.P.T. adversary $\mathcal{A}$: $\mathcal{A}$ can use oracle O_{cu} to get user secret/public key pairs, and use $(O_{pke},\ O_{pkd})$ for the CCA attack game queries. Then $\mathcal{A}$ generates $(upk_{B_0}, \hat{mpk}_{0_0}), (upk_{B_1}, \hat{mpk}_{0_1})$ sends them to the challenger. The challenger computes $req^* \leftarrow \text{Enc}((upk_{B_b}, \hat{mpk}_{0_b}), upk_a)$ where $b \leftarrow \{0,1\}$, and sends it to $\mathcal{A}$, the adversary then outputs $\hat{b} \in \{0,1\}$. We define Exp_{req_b} to be the adversary outputs 1 in experiment b, then the advantage of $\mathcal{A}$ in User query security is denoted as $Adv_{\mathcal{A}}^{\text{UQS}}$.

Exp_{req_b}:

1 : $b \leftarrow\!\!\$\ \{0, 1\}$

2 : $(upk_a, usk_a) \leftarrow\!\!\$\ \text{PKeyGen}()$

3 : $((upk_{B_0}, \hat{mpk}_{B_0}), (upk_{B_1}, \hat{mpk}_{B_1})) \leftarrow\!\!\$\ \mathcal{A}(upk_a)$

4 : $req^* \leftarrow\!\!\$\ \text{Enc}(upk_a, (upk_{B_b}, \hat{mpk}_{B_b}))$

5 : $b' \leftarrow\!\!\$\ \mathcal{A}((upk_a, req^*)$

6 : **return** $b = b'$

Fig. 6. Experiments of User query security

Definition 7. *(User query security). Our scheme provides user query security if for any P.P.T. adversary $\mathcal{A}$:*

$$Adv_{\mathcal{A}}^{\text{UQS}} = |Pr[Exp_{req_0}] - Pr[Exp_{req_1}]|\ \text{is negligible.}$$

Theorem 2. *Our scheme provides user request security if the public encryption scheme we applied here satisfies CCA indistinguishability.*

Our scheme requires an unforgeable cross-chain consensus, where the adversary is unable to forge a cross-chain consensus with an aggregate signature that can be verified and decrypted correctly.

Consider the experiment below: For a P.P.T. adversary $\mathcal{A}$, $\mathcal{A}$ can use oracle O_{cm} to get any consortium member secret/public key pair, and use oracle $O_{sig}(m)$ to get the signature of m (We request that the oracle O_{cm} can be used at most $\frac{2}{3}N - 1$ times). The adversary is able to interact with the consortium blockchain to obtain $ccs_i, i \in [1, Q]$ (Q is the maximum number that $\mathcal{A}$ can query to the challenger). Given ccs_i, the challenger computes $m_i \leftarrow \text{EncDec}(ccs_i, ask)$

and sends m_i to $\mathcal{A}$. Finally $\mathcal{A}$ computes a valid forgery pair (css^*, m^*) which is not in the set (ccs_i, m_i), $i \in [1, Q]$ and this forgery pair can be successfully verified by the trust agent and received by the user. We denote the advantage of $\mathcal{A}$ to generate a valid pair (css^*, m^*) as $Adv_{\mathcal{A}}^{CCU}$.

Definition 8. *(Cross-chain consensus unforgeability). Our scheme satisfies cross-chain consensus unforgeability if for any P.P.T. adversary $\mathcal{A}$, $Adv_{\mathcal{A}}^{CCU}$ is negligible.*

Theorem 3. *Our scheme provides Cross-chain consensus unforgeability if the aggregate signature scheme we applied here satisfies unforgeability.*

Here we consider user information security, to be more precise, attackers are unable to derive m from css and $cssu$. For a P.P.T. adversary $\mathcal{A}_0$: the adversary generates $(tx_0, m_0), (tx_1, m_1)$ and sends them to the challenger, the challenger computes $css^* \leftarrow$ CreateCss $((tx_b, m_b), msk)$ where $b \leftarrow \{0, 1\}$, and sends it to $\mathcal{A}_0$, the adversary outputs $\hat{b} \in \{0, 1\}$. We define Exp_{css_b} to be the adversary outputs 1 in experiment b, then the advantage of $\mathcal{A}_0$ in consortium response security is denoted as:

$$Adv_{\mathcal{A}}^{CRS} = |Pr[Exp_{css_0}] - Pr[Exp_{css_1}]|$$

Similarly, for a P.P.T. adversary $\mathcal{A}_1$: the adversary generates $(tx_0, m_0, k_0, \hat{mpk}_0), (tx_1, m_1, k_1, \hat{mpk}_1)$ and sends to the challenger, the challenger computes $cssu^* \leftarrow$ CreateCssU $((tx_b, m_b, k_b, \hat{mpk}_b), msk)$ where $b \leftarrow \{0, 1\}$, and sends it to $\mathcal{A}_1$, the adversary outputs $\hat{b} \in \{0, 1\}$. Define Exp_{cssu_b} to be the adversary outputs 1 in experiment b, then the advantage of $\mathcal{A}_1$ in user response security is denoted as:

$$Adv_{\mathcal{A}}^{URS} = |Pr[Exp_{cssu_0}] - Pr[Exp_{cssu_1}]|$$

We denote the advantage of adversary in User information security as $Adv_{\mathcal{A}}^{UIP}$ (Fig. 7).

Definition 9. *(User information security). Our scheme provides user information security if for a P.P.T. adversary $\mathcal{A}_0$: $Adv_{\mathcal{A}_0}^{CRS}$ is negligible, for a P.P.T. adversary $\mathcal{A}_1$: $Adv_{\mathcal{A}_1}^{URS}$ is negligible, then for any P.P.T. adversary $\mathcal{A}$: $Adv_{\mathcal{A}}^{UIP}$ is negligible.*

Theorem 4. *Our scheme satisfies User information security if the public encryption scheme we applied here satisfies indistinguishability, and cross-chain messages css and cssu leak no information about m.*

Proof. The property of user information privacy for $cssu$ is similar to the indistinguishability of the public encryption scheme. The property of user information privacy for css can be directly derived from $EncProof$ indistinguishable from

Exp_{css_b}:	Exp_{cssu_b}:
1 : $b \leftarrow \$ \{0,1\}$	1 : $b \leftarrow \$ \{0,1\}$
2 : $(mpk, msk) \leftarrow \$ \text{CMKeyGen}()$	2 : $(mpk, msk) \leftarrow \$ \text{CMKeyGen}()$
3 : $((tx_0, m_0), (tx_1, m_1)) \leftarrow \$ \mathcal{A}(mpk)$	3 : $((tx_0, m_0, k_0, \hat{mpk}_0), (tx_1, m_1, k_1, \hat{mpk}_1)) \leftarrow \$ \mathcal{A}(mpk)$
4 : $css^* \leftarrow \$ \text{CreateCss}((tx_b, m_b), msk)$	4 : $cssu^* \leftarrow \$ \text{CreateCssU}((tx_b, m_b, k_b, \hat{mpk}_b), msk)$
5 : $b' \leftarrow \$ \mathcal{A}((mpk, css^*)$	5 : $b' \leftarrow \$ \mathcal{A}((mpk, cssu^*)$
6 : **return** $b = b'$	6 : **return** $b = b'$

Fig. 7. Experiments of User information security

Definition 2 and *Transaction indistinguishable* from Definition 5. If a P.P.T. adversary $\mathcal{A}_1$ can break the indistinguishability of css, then it is able to break *Transaction indistinguishable* as well. Thus both css and $cssu$ leak no user information, and our scheme satisfies user information privacy.

We now focus on the ability of the auditor: All css that can be verified correctly on the consortium blockchain can only be decrypted by the auditor to obtain m. Anyone except the auditor is unable to decrypt css on the consortium blockchain. Then, the auditability requires consortium responses with cross-chain consensus to be successfully verified and decrypted. We define the advantage of the adversary to obtain m under the conditions mentioned above as $Adv_{\mathcal{A}}^{ADT}$.

Definition 10. *(Auditability). Our scheme satisfies auditability if for a P.P.T. adversary* $\mathcal{A}_0$: $Adv_{\mathcal{A}_0}^{CCU}$ *is negligible, for a P.P.T. adversary* $\mathcal{A}_1$: $Adv_{\mathcal{A}_1}^{UIP}$ *is negligible, then for any P.P.T. adversary* $\mathcal{A}$: $Adv_{\mathcal{A}}^{ADT}$ *is negligible.*

Theorem 5. *Our scheme satisfies auditability if it also satisfies Cross-chain consensus unforgeability and User information security.*

6 Features, Efficiency and Applications

6.1 Features and Efficiency

Our solution implements the following features:

Cross-Chain Data Transfer: We achieve cross-chain data transmission through the consortium blockchain to the public blockchain. The scheme enables the transfer of both public information and personal private information by using a trusted agent, public information and user private information can be sent from the consortium blockchain to designated public nodes.

Unforgeability: Malicious consortium members are unable to generate forgery user information and submit them to the consortium blockchain. All consensus algorithms are effective and can be read by the trust agent and submitted to the public blockchain.

Anonymity: For certain specific scenarios, both parties involved in a matching transactions unwilling to disclose their identities to the public. It is achieved by utilizing confidential transactions and compatible signatures, replacing the key corresponding to the identity with a special commitment to realize the verification of information by different parties, as well as identity privacy for both sending and receiving parties.

User Information Privacy: In certain specific scenarios, for a user the information on the public blockchain ensures that no one else can use it for matching. More simply, the user private information for public matching is only known by the consortium member which generates it, two match parties and the auditor.

Auditability: The auditor can decrypt a transaction in the consortium blockchain to recover transaction with necessary, it can reveal the detail of the transaction and each consortium member which signed the transaction to enclose a malicious attacker, the auditor can reveal the member who submit the transaction to the consortium blockchain.

Our solution helps reduce blockchain storage costs to a certain extent:

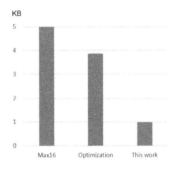

Fig. 8. Comparison of 32-bits range proofs cost

In blockchain transactions, range proofs consume a significant amount of storage in blockchain, leading to excessively large ledger information. As depicted in Fig. 8, our approach, compared with current implementations such as [17], effectively reduces the size of the range proof from $O(n)$ to $O(\log(n))$. This reduction results in a range proof cost of approximately 1 KB for 32-bit messages, a substantial improvement compared to 5.4 KB and 3.8 KB with optimizations.

6.2   Applications

Our solution facilitates data transfer from a consortium blockchain to a public blockchain using cross-chain technology. With support for vector input, our scheme finds straightforward applications in scenarios involving control signal transmission and anonymous notarization. The aggregation transaction function in our scheme has the potential to significantly reduce the expense of cross-chain transmission. In notarization scenarios, multiple requests from users on the same node can be efficiently handled through aggregated signatures, streamlining the information transmission process.

Furthermore, our solution has the potential to be applied in the exchange and matching of educational and medical information, especially within educational settings. Educational institutions can take on the role of consortium blockchain members, and these members can release information in a quantitative format. Users' academic performance, educational information, and personal abilities can all be shared by consortium blockchain members for public chain matching.

In medical scenarios, members of the consortium blockchain can include diverse medical institutions, while patients, as users, initiate queries for medical information through trusted agents. Various parameters in the medical process can serve as inputs for transactions, and patients can utilize medical data through public blockchain matching. These methods effectively ensure the privacy protection of sensitive medical and identity information exchanged between medical institutions and patients.

7   Conclusion

In this paper, we propose an information matching scheme based on cross-chain technology. Unlike existing blockchain systems, information platforms, and traditional information matching systems, our scheme facilitates consensus between different blockchains, offering features such as unforgeability, anonymity, information privacy, and cross-chain information auditability. Our scheme functions as a distributed information transmission system capable of auditing public transactions created by consortium members to reveal the details of user information involved in an information matching process. Furthermore, our scheme demonstrates resistance within a specific threat model and satisfies various security properties under this model.

References

1. Nakamoto, S.: Bitcoin: a peer-to-peer electronic cash system. Decentral. Bus. Rev. (2008)
2. Wüst, K., Kostiainen, K., Capkun, V., Capkun, S.: Prcash: centrally-issued digital currency with privacy and regulation. IACR Cryptology ePrint Archive 2018, p. 412 (2018)
3. Noura, M., Atiquzzaman, M., Gaedke, M.: Interoperability in internet of things: taxonomies and open challenges. Mob. Netw. Appl. **24**, 796–809 (2019)
4. Celer homepage. https://celer.network/. Accessed 10 Nov 2023
5. Multichain homepage. https://multichain.org. Accessed 10 Nov 2023
6. Synapse homepage. https://www.synapseprotocol.com. Accessed 10 Nov 2023
7. Umbri homepage. https://bridge.umbria.network/. Accessed 10 Nov 2023
8. Ghosh, B.C., Bhartia, T., Addya, S.K., Chakraborty, S.: Leveraging public-private blockchain interoperability for closed consortium interfacing. In: IEEE INFOCOM 2021-IEEE Conference on Computer Communications, pp. 1–10. IEEE (2021)

9. Syta, E., et al.: Keeping authorities "honest or bust" with decentralized witness cosigning. In: 2016 IEEE Symposium on Security and Privacy (SP), pp. 526–545. IEEE (2016)

10. Castro, M., Liskov, B.: Practical byzantine fault tolerance and proactive recovery. ACM Trans. Comput. Syst. (TOCS) **20**(4), 398–461 (2002)

11. Garoffolo, A., Kaidalov, D., Oliynykov, R.: Zendoo: A zk-snark verifiable cross-chain transfer protocol enabling decoupled and decentralized sidechains. In: 2020 IEEE 40th International Conference on Distributed Computing Systems (ICDCS), pp. 1257–1262. IEEE (2020)

12. Yin, Z., Zhang, B., Xu, J., Lu, K., Ren, K.: Bool network: an open, distributed, secure cross-chain notary platform. IEEE Trans. Inf. Forensics Secur. **17**, 3465–3478 (2022)

13. He, Y., Zhang, C., Wu, B., Yang, Y., Xiao, K., Li, H.: Cross-chain trusted service quality computing scheme for multi-chain model-based 5g network slicing SLA. IEEE Internet Things J. (2021)

14. Yi, L., et al.: CCUBI: a cross-chain based premium competition scheme with privacy preservation for usage-based insurance. Int. J. Intell. Syst. **37**(12), 11522–11546 (2022)

15. Chen, K., Lee, L.F., Chiu, W., Su, C., Yeh, K.H., Chao, H.C.: A trusted reputation management scheme for cross-chain transactions. Sensors **23**(13), 6033 (2023)

16. Pedersen, T.P.: Non-interactive and information-theoretic secure verifiable secret sharing. In: Feigenbaum, J. (ed.) CRYPTO 1991. LNCS, vol. 576, pp. 129–140. Springer, Heidelberg (1992). https://doi.org/10.1007/3-540-46766-1_9

17. Maxwell, G.: Confidential transactions (2015) (2016)

18. Boneh, D., Boyen, X.: Short signatures without random oracles. In: Cachin, C., Camenisch, J.L. (eds.) EUROCRYPT 2004. LNCS, vol. 3027, pp. 56–73. Springer, Heidelberg (2004). https://doi.org/10.1007/978-3-540-24676-3_4

19. Boneh, D., Lynn, B., Shacham, H.: Short signatures from the Weil pairing. In: Boyd, C. (ed.) ASIACRYPT 2001. LNCS, vol. 2248, pp. 514–532. Springer, Heidelberg (2001). https://doi.org/10.1007/3-540-45682-1_30

20. Boneh, D., Gentry, C., Lynn, B., Shacham, H.: Aggregate and verifiably encrypted signatures from bilinear maps. In: Biham, E. (ed.) EUROCRYPT 2003. LNCS, vol. 2656, pp. 416–432. Springer, Heidelberg (2003). https://doi.org/10.1007/3-540-39200-9_26

21. Camenisch, J., Chaabouni, R., shelat, A.: Efficient protocols for set membership and range proofs. In: Pieprzyk, J. (ed.) ASIACRYPT 2008. LNCS, vol. 5350, pp. 234–252. Springer, Heidelberg (2008). https://doi.org/10.1007/978-3-540-89255-7_15

22. Bünz, B., Bootle, J., Boneh, D., Poelstra, A., Wuille, P., Maxwell, G.: Bulletproofs: short proofs for confidential transactions and more. In: 2018 IEEE Symposium on Security and Privacy (SP), pp. 315–334. IEEE (2018)

23. Androulaki, E., et al.: Hyperledger fabric: a distributed operating system for permissioned blockchains. In: Proceedings of the Thirteenth EuroSys Conference, pp. 1–15 (2018)

24. Fuchsbauer, G., Orrù, M., Seurin, Y.: Aggregate cash systems: a cryptographic investigation of mimblewimble. In: Ishai, Y., Rijmen, V. (eds.) EUROCRYPT 2019. LNCS, vol. 11476, pp. 657–689. Springer, Cham (2019). https://doi.org/10.1007/978-3-030-17653-2_22

25. Yuen, T.H.: PAChain: private, authenticated and auditable consortium blockchain. In: Mu, Y., Deng, R.H., Huang, X. (eds.) CANS 2019. LNCS, vol. 11829, pp. 214–234. Springer, Cham (2019). https://doi.org/10.1007/978-3-030-31578-8_12

Verifiable Attribute-Based Proxy Re-encryption with Non-repudiation Based on Blockchain

Yaorui He[1], Ting Liang[1], Pei Huang[1], and Zhe Xia[1,2(✉)]

[1] School of Computer Science and Artificial Intelligence, Wuhan University of Technology, Wuhan 430070, China
xiazhe@whut.edu.cn
[2] Hubei Key Laboratory of Transportation Internet of Things, Wuhan University of Technology, Wuhan 430071, China

Abstract. With the development of Blockchain, cloud computing, and artificial intelligence, smart transportation is highly likely to drive urban transportation in the direction of intelligence and digitalization. However, when connected vehicle devices share data in edge networks, there are security and privacy issues, e.g. leakage of sensitive information will cause serious threats. The separation of user data ownership and physical control requires users to impose access control on the outsourced data. Existing privacy protection schemes still suffer from problems such as a lack of supervision of centralized third-party cloud servers, which face the challenge of data leakage. In this paper, Blockchain and proxy re-encryption techniques are used to address these challenges. To enable sharing of the outsourced data in a secure way, Attribute-Based Proxy Re-encryption (ABPRE) is a popular technique. At the same time, verifiability enables the shared user to verify that the re-encrypted ciphertext returned by the server is correct. In addition, the introduction of a decentralized Blockchain provides distributed storage and it is tamper-proof for enhancing the trustworthiness of the connected vehicle devices as well as the security of the communication between the entities. The security and performance analyses demonstrate that our proposed scheme can achieve the desirable security features and it is efficient for protecting data privacy.

Keywords: Blockchain · Proxy Re-Encryption · Privacy-Preserving Data-Sharing

1 Introduction

The Intelligent Transportation System(ITS) [15] is a new information technology that integrates the Internet of Things (IoT), spatial awareness, cloud computing, and mobile Internet in the field of transportation. The ITS structure can be divided into three layers: the perception layer, the network layer, and the

J. Chen and Z. Xia (Eds.), BlockTEA 2023, LNICST 577, pp. 115–134, 2024.
https://doi.org/10.1007/978-3-031-60037-1_7

application layer. Among them, the perception layer includes vehicle nodes and the sensors configured around them, such as GPS and cameras, to collect vehicle driving data in real-time. The network layer is responsible for transmitting information and realizing data transmission and sharing between entities such as vehicle-to-vehicle (V2V) and vehicle-to-infrastructure(V2I) through wireless communications such as Zigbee, WiFi, and DSRC [13]. Service providers collect vehicle data and analyze and process them through cloud computing technology, and provide services to users at the application layer to improve the traffic environment and urban operation efficiency.

Data sharing and cooperation are key to achieving smart transportation and the Internet of Vehicles(IoV). Communication networks between vehicles (e.g. V2V) can share information such as real-time traffic conditions, vehicle location, and speed. Through direct data sharing, vehicles understand each other's status, helping to optimize route selection, avoid congestion, and provide early warning and emergency response, among other things [3]. Data sharing between vehicle manufacturers and related service providers allows third-party developers and applications to access and utilize vehicle data, facilitating innovation and the development of various connected vehicle applications such as navigation software, vehicle-sharing services, etc. Open data interfaces can follow standardized data formats and protocols to ensure data interoperability and security [18]. In general, participants can enter into data cooperation and sharing agreements that specify how data is to be shared permissions, and privacy protection [2]. This can involve cooperation between vehicle manufacturers, transportation authorities, city planners, insurance companies, and other parties. These data-sharing and cooperation methods can be achieved through various technical means, such as V2V, V2I, IoT technologies, cloud computing, etc.

While cloud-assisted IoV data sharing offers many conveniences and innovative features, it also raises a number of security and privacy concerns. Data from vehicles is transferred to cloud servers for processing and storage, which leads to having to take the risks associated with data centralization. In the event of an attack or failure of the cloud server, a large amount of vehicle data may be lost or unavailable [12]. Also, vehicle data contains personal sensitive information such as driving habits and destination information of the vehicle owner, which is at risk of data leakage, data tampering, or unauthorized access [21]. With the separation of data ownership and storage, data owners have a strong incentive to maintain control over their use and access to shared data.

To address these issues, a number of security and privacy protection measures are required, such as

- Data privacy: The vehicle data is encrypted before uploading to the cloud to ensure data privacy.
- Access control: Restrict access to vehicle data by other entities and authorize only to trusted service providers and authorized users.
- Compliance review: Ensure that data is shared in compliance with applicable laws and regulations and that privacy protection requirements are observed.

– Security audits: Audits and vulnerability scans for cloud-assisted Internet of Vehicles security to identify and fix potential security issues in a timely manner.

In this paper, Blockchain and proxy re-encryption technologies are integrated to address the above privacy and security problems. Blockchain is a decentralized ledger that ensures security and trustworthiness of data through cryptographic means. Blockchain uses techniques such as public key encryption and hash functions to verify and protect the integrity of data, while consistency of the ledger is ensured by consensus algorithms. Blockchain is best known in the field of cryptocurrencies, such as Bitcoin [16] and Ether, but also has many other application areas, such as supply chain management, smart contracts, etc. Blockchain and the Internet of Vehicles are two different but widely used technology areas. Their technological convergence can bring many advantages to the intelligent transportation system and automotive industry [5]. Proxy Re-Encryption [1] is a novel technique to allow a third party to transform some encrypted data from one public key to another public key without learning the underneathing plaintext. In IoV, attribute-based proxy re-encryption schemes can help enable secure and privacy-preserving data sharing. However, existing privacy protection schemes still suffer from problems such as a lack of supervision of centralized third-party cloud servers [9]. Since ciphertext processing consumes cloud computing resources, some inactive cloud servers may even return the wrong ciphertext to save computing resources.

The integration of Blockchain and proxy re-encryption can enhance data protection and access control. Access and sharing of data on the Blockchain is controlled through the use of proxy re-encryption, allowing only specific authorized users to decrypt and access the data. At the same time, the Blockchain ensures that the data stored by the user is not tampered with. This combination provides a more robust permission management and privacy protection mechanism, while still retaining the decentralized and trustworthy characteristics of the Blockchain.

The main contributions of this paper are as follows.

1. First of all, this paper adopts a privacy protection scheme based on proxy re-encryption to realize the data sharing of the Internet of Vehicles, which can simplify the sharing process between entities, and users do not need to download the ciphertext to decrypt it before re-encrypting and uploading it.
2. Secondly, the scheme proposed in this paper provides verifiability, which means that users can directly verify the legitimacy of the re-encrypted ciphertext without having to perform complex calculations.
3. Finally, this paper combines Blockchain technology to achieve persistent storage, provide evidence for arbitration, and solve the problems that centralized cloud servers are vulnerable to single-point attacks and data tampering.

The remainder of the paper is organized as follows: Sect. 2 reviews the state of the art of relevant research addressing the problem considered in this paper. Section 3 presents the cryptographic primitives and basic definitions involved in implementing the scheme proposed in this paper. Section 4 states the various models and definitions involved in the scheme of this paper. Section 5 shows the details of the proposed scheme and the protocol flow. Section 6 proves the security of the scheme in this paper. Section 7 provides a comparative analysis of the schemes related to this paper. Section 8 concludes the paper and provides an outlook on future developments.

2   Related Work

Proxy Re-Encryption allows a third-party, called the proxy server, to transform ciphertexts without revealing the original key [4]. This technique can be used to securely share encrypted data and has important applications in privacy protection and data security.

Proxy re-encryption techniques can realize a variety of functions, such as transformation and decryption of proxy re-encryption, delegation, and revocation of proxy re-encryption. Various proxy re-encryption schemes have been proposed, including identity-based schemes [8,10,19], certificate-less schemes [20], and attribute-based schemes [14] etc. Researchers endeavor to propose proxy re-encryption schemes with strong security guarantees and verify them with formal security proofs [11]. In order to detect whether the proxy is malicious, Ohata et al. [17] proposed a mechanism to verify the output of the proxy, but it has not considered fine-grained access control. Liang et al. [14] introduced the concept of proxy re-encryption into attribute-based encryption schemes, thereby achieving flexible control over encrypted data. Ge et al. [9] proposed a verifiable and fair ciphertext-policy attribute-based proxy re-encryption scheme to prevent malicious cloud servers from forging ciphertext as their malicious behavior can be identified. PRE is widely used in cloud computing, data sharing, and multi-party secure computing. Researchers are committed to improving the efficiency, security, and functionality of proxy re-encryption to meet the needs of different application scenarios.

Overall, proxy re-encryption techniques are an active research area, and researchers are continuously improving and enhancing their efficiency and security. With the increased demand for data security and privacy protection, proxy re-encryption techniques are widely used in more application scenarios.

3   Preliminaries

This section introduces the cryptographic primitives needed to construct our scheme as well as the Blockchain basics to make it easier for the reader to understand our work.

3.1   Attribute-Based Proxy Re-Encryption

A proxy re-encryption scheme is a cryptographic technique used to implement the transformation and re-encryption of ciphertext without exposing the original key. This technique can be used to securely share encrypted data and has important applications in privacy protection and data security. The following are the basic elements of the proxy re-encryption scheme.

The PRE scheme involves multiple entities, including:

- delegator: It is the data owner that encrypts the plaintext using the public key.
- proxy: It is an intermediate third party that performs the re-encryption operation to convert the ciphertext from the delegator's public key to the delegatee's public key.
- delegatee: It is the intended recipient of the message, who has the corresponding private key to decrypt the re-encrypted ciphertext.

An ABPRE scheme is a tuple of PPT algorithms (SETUP, KEYGEN, RKGEN, ENC, REENC, DEC).

1. $SETUP(1^k) \rightarrow (params, mk)$: takes as inputs a security parameter 1^k, and it outputs the public parameter $params$ and the master key mk.
2. $KEYGEN(S, mk) \rightarrow (usk)$: takes as inputs an attribute set S and the master key mk, and it outputs a secret key usk.
3. $ENC(AS, m) \rightarrow (C)$: takes as inputs an access structure AS and a message m, and it outputs a ciphertext C.
4. $RKGEN(usk, AS) \rightarrow (rk)$: takes as inputs a secret key usk and an access structure AS, and it outputs a re-key rk.
5. $REENC(rk, C) \rightarrow (C')$: takes as inputs a re-key rk and a ciphertext C, and it checks whether the access structure of C is satisfied by the attribute set in rk. If yes, it outputs a re-encrypted ciphertext C', or it outputs "reject" otherwise.
6. $DEC(usk, C) \rightarrow (m)$: takes as inputs a secret key usk and a ciphertext C, and it checks if the access structure of C is satisfied by the attribute set in usk. If yes, it outputs a message m in the message space, or it outputs "reject" otherwise.

Correctness. This property captures two requirements, i.e. for any message m in the message space, the following two equations should hold:

1. $DEC(usk_S, ENC(AS, m)) = m$;
2. $DEC(usk_{S'}, REENC(rk_{AS \rightarrow AS'}, C)) = m$.

where S satisfies AS, S' satisfies AS', mk is the master key, C is the ciphertext associate with the message m and the access structure AS'.

3.2 Blockchain

Blockchain technology is a distributed ledger, and its main features are as follows:

1. Distributed ledger: Blockchain is a decentralized ledger that is maintained and shared by multiple entities. Each has a complete copy of all transaction records and block information. Distributed ledgers are characterized by decentralization, openness and transparency, security, and trust.

2. Block and chain: the block is a basic unit in a Blockchain, containing a series of transaction records as well as some metadata. Each block contains a hash value that points to the previous block, forming a chain structure. And this is why it is called Blockchain. New blocks are inserted to the end of the chain by a specific consensus algorithm and its main feature is tamper-proof.
3. Distributed Consensus: Blockchain solves the problem of consistency among participants using distributed consensus algorithms. The consensus algorithm ensures that all participants agree on the state of the ledger without a trusted authority. Proof of Work (PoW) and Proof of Stake (PoS) are commonly used consensus algorithms.
4. Cryptography: Blockchain uses cryptographic techniques, such as digital signature and hash function, to ensure the security and privacy protection of data. Hash functions are used to generate unique identifiers for blocks, and digital signatures are used to prove the legitimacy of transactions.
5. Decentralized applications: Blockchain can support decentralized applications (DApps), which are applications built on the Blockchain that perform various functions, such as digital asset trading, smart contract execution, data storage, validation, etc. Decentralized applications enable trust and security without the need for intermediaries.
6. Smart Contracts: They are automated contracts executed on the Blockchain according to some predefined rules and conditions. Smart contracts can automatically execute and validate transactions without a trusted third party. They can also implement complex logic such as asset transfers, conditional payments, voting, etc.

The practical applications of Blockchain technology are very wide, covering a wide range of fields such as smart transportation, smart healthcare, and industrial Internet of Things. With the development and innovation of technology, Blockchain will continue to drive the digital economy and social change.

4   Models and Definitions

4.1   System Model

A verifiable attribute-based proxy re-encryption system includes the following entities, key generates center, data owner, original recipient, a cloud server that acts as a proxy, and a shared user. As shown in Fig. 1.

– Key Generation Center: KGC is responsible for establishing the cryptosystem, publishing the system's public parameters, and saving the master key. It accepts user (DO, DR, SU) registration requests, generates user private keys, and sends them to users via secure channels.
– Data Owners: DO is responsible for collecting plaintext data, which needs to be encrypted to protect user privacy before uploading to the Blockchain.
– Data Recipient: DR can access the ciphertext on the Blockchain and it can use its private key to recover the plaintext. Moreover, DR can share the ciphertext with other users.

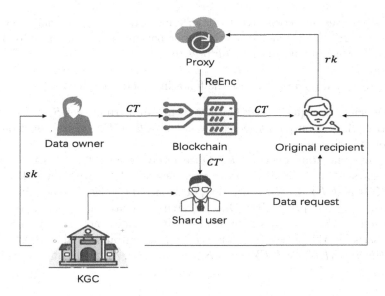

Fig. 1. System model

- Proxy: The Proxy is responsible for handling data-sharing requests. The DR generates the re-encryption key, which is used by the Proxy to re-encrypt the original ciphertext and re-upload the processing result to the Blockchain.
- Shared User: SU can request access to the ciphertext on the Blockchain, first he needs to make an access request to the original data recipient, and after being authorized, he can download the re-encrypted ciphertext from the Blockchain and decrypt it with his own private key.

4.2 Threat Model

In our threat model, the DR and SU are both assumed to be honest-but-curious.

- An external adversary without a valid key, including a proxy, attempts to obtain information through the ciphertext data on the Blockchain.
- Proxy and unauthorized shared users may conspire to attempt to obtain other data information, even private keys, from the original recipient.
- Proxy may reuse previous re-encrypted ciphertexts, or randomly select re-encrypted ciphertexts to save costs.

4.3 Security Requirements

The following security requirements are considered in our proposed scheme.

Selective-Structure Chosen Plaintext Security. An ABPRE scheme is said to be secure against selective-structure chosen plaintext attack, if no PPT adversary $\mathcal{A}$ can win the SS-CPA game with non-negligible advantage.

Init: $\mathcal{A}$ chooses an access structure AS^* and sends it to the challenger, who then runs $SETUP(1^k)$ and sends the public parameter *params* back to $\mathcal{A}$. The challenger keeps the master key mk private.

Phase 1: The adversary $\mathcal{A}$ issues the following queries to the oracles:

- Key generation query $\mathcal{O}kg$: Once receiving an attribute set S, if S does not satisfy AS^*, it outputs a secret key $usk = KEYGEN(S, mk)$, or it outputs "reject" otherwise.
- Re-key generation query $\mathcal{O}rkg$: Once receiving an attribute set S and an access structure AS, if S does not satisfy AS^*, it outputs a re-key $rk = RKGEN(KEYGEN(S, mk), AS)$, or it outputs "reject" otherwise.
- Re-encryption query $\mathcal{O}re$: Once receiving an attribute set S, an access structure AS and a ciphertext C, if S does not satisfy AS^* and S satisfies the access structure of C, it outputs a ciphertext $C' = REENC(RKGEN(KEYGEN(S, mk), AS), C)$, or it outputs "reject" otherwise.

Challenge: When the above phase is over, $\mathcal{A}$ outputs two messages m_0, m_1 with equal length. The challenger randomly chooses $b \in \{0, 1\}$ and encrypts m_b with AS^*. Then, the ciphertext C^* is sent to $\mathcal{A}$.

Phase 2: The same as in Phase 1.

Guess: $\mathcal{A}$ guesses a bit $b' \in \{0, 1\}$, and it wins the game if $b' = b$.
$\mathcal{A}$'s advantage in SS-CPA game is defined as follows:

$$Adv_{SS-CPA,\mathcal{A}}^{\mathcal{O}kg,\mathcal{O}rkg,\mathcal{O}ree} = |Pr[b' = b] - 1/2| \tag{1}$$

Key Security for ABPRE. An ABPRE scheme enjoys the key security property if no PPT adversary $\mathcal{A}$ can win the following MKS game with a non-negligible advantage.

Init: $\mathcal{A}$ randomly chooses an attribute set S^* for the challenge and sends it to the challenger, who then runs $SETUP(1^k)$ and sends the public parameters *params* back to $\mathcal{A}$. The challenger keeps the master-key mk private.

Queries: $\mathcal{A}$ issues the following queries to the oracles:

- Key generation query $\mathcal{O}kg$: Once receiving an attribute set S, if $S \neq S^*$, it outputs a secret key $usk = KEYGEN(S, mk)$, or it outputs "reject" otherwise.
- Re-key generation query $\mathcal{O}rkg$: Once receiving an attribute set S and an access structure AS, it outputs $rk = RKGEN(KEYGEN(S, mk), AS)$.

Output: Note that re-encryption query and decryption query are straightforward, because the re-key can be generated by querying the re-key generation oracle. Finally, $\mathcal{A}$ outputs the secret key usk^* for attribute set $S*$.

$\mathcal{A}$'s advantage for the above MKS game is defined as

$$Adv_{MKS,\mathcal{A}}^{\mathcal{O}kg,\mathcal{O}rkg} = |Pr[\mathcal{A} \ succceeds]| \tag{2}$$

Verifiability. An ABPRE scheme satisfies the verifiability property if no PPT adversary $\mathcal{A}$ can win the Ver game with a non-negligible advantage.

Init: The challenger runs $SETUP(1^k)$ and gives $\mathcal{A}$ the system parameters *params*. It keeps the corresponding master-key mk to itself.

Phase 1: The adversary $\mathcal{A}$ issues queries to the oracles:

- Key generation query $\mathcal{O}kg$: Once receiving an attribute set S, it outputs a secret key $usk = KEYGEN(S, mk)$.
- Re-key generation query $\mathcal{O}rkg$: Once receiving an attribute set S and an access structure AS, it outputs a re-key $rk = RKGEN(KEYGEN(S, mk), AS)$.
- Claim query $\mathcal{O}claim$: Suppose the claim query is issued on (CT, rk, CT'). The challenger outputs $CLAIM(CT, rk, CT')$.

Challenge: $\mathcal{A}$ chooses a message m^* and some access policy (AS^*, ρ^*), and sends them to challenger, who then computes $CT^* = Enc(m^*; (AS^*; \rho^*))$ and sends CT^* back to $\mathcal{A}$.

Phase 2: The same as in Phase 1.

Guess: $\mathcal{A}$ outputs an attribute set S^* as well as a re-encrypted ciphertext CT'^*, where $R(S^*; (AS^*; \rho^*)) = 1$. $\mathcal{A}$ wins the game if $Dec_{re}(AS^*; CT^*; CT'^*) \notin \{m^*, \bot\}$.

$\mathcal{A}$'s advantage of winning the above game is defined as

$$Adv_{\mathcal{A}}^{Ver} = |Pr[\mathcal{A} \ succceeds]| \tag{3}$$

4.4   Assumptions

Bilinear Map. Let G_1 and G_2 be two multiplicative cyclic groups of prime order p. Let g be a generator of G_1 and e be a bilinear map, $e : G_1 \times G_1 \to G_2$. The bilinear map e has the following properties:

1. Bilinearity: If for all $u, v \in G_1$ and $a, b \in Z_p$, we have $e(u^a, v^b) = e(u, v)^{ab}$, Or $e(u \cdot v, w) = e(u, w) \cdot e(v, w)$ and $e(u, v \cdot w) = e(u, v) \cdot e(u, w)$, then the mapping is said to be bilinear.

2. Non-degeneracy: The mapping does not map all pairs of elements in $G_1 \times G_1$ to unit elements in G_2. Since G_1, and G_2 are all groups of order prime, this implies that: if g is the generating element of G_1, then $e(g, g)$ is the generating element of G_2.
3. Computability: For any $u, v \in G_1$, an efficient algorithm exists to compute $e(u, v)$.

Decisional Parallel Bilinear Diffie-Hellman Exponent Assumption. Denote (e, G_1, G_2, p, g) as a bilinear pairing, randomly choose $a, s, b_1, ..., b_q \in Z_p$ and g is a generator of G_1. Denote y as

$$g, g^s, g^a, ..., g^{(a^q)}, g^{(a^{q+2})}, ..., g^{(a^{2q})},$$

$$\forall_{1 \leq j \leq q}, g^{s \cdot b_j}, g^{a/b_j}, ..., g^{(a^q/b_j)}, g^{(a^{q+2}/b_j)}, g^{(a^{2q}/b_j)},$$

$$\forall_{1 \leq j, k \leq q, k \neq j}, g^{a \cdot s \cdot b_k/b_j}, g^{(a^q \cdot s \cdot b_k/b_j)}$$

the q-parallel BDHE assumption requires that for any PPT adversary $\mathcal{A}$, its advantage of distinguishing $e(g, g)^{a^{q+1} \cdot s} \in G_2$ from a random element in G_2 is negligible. Formally, the advantage of $\mathcal{A}$ is

$$Adv^{BDHE} = |Pr[\mathcal{B}(y, T = e(g, g)^{a^{q+1} \cdot s}) = 0] - Pr[\mathcal{B}(y, T = R) = 0]|.$$

Discrete Logarithm Assumption. Denote (e, G_1, G_2, p, g) as a bilinear pairing, for a tuple $(e, G_1, G_2, p, g, g^\beta)$ where $g \in G_1, \beta \in Z_p^*$, the discrete logarithm assumption requires that for any PPT adversary $\mathcal{A}$, its advantage of computing the integer β is negligible. Formally, the advantage of $\mathcal{A}$ is

$$Adv^{DL} = Pr[\mathcal{A}(e, G_1, G_2, p, g, g^\beta) = \beta].$$

5 Proposal Scheme

In this section, we first introduce a verifiable attribute-based proxy re-encryption scheme, which mainly consists of eight algorithms. First, the authorization center (KGC) generates the public parameters and keeps the master key secret. Participants then register (DO, DR, SU) with the authorization center. This requires these participants to authenticate their identities, and then they can obtain the private keys associated with their attributes. The data owner collects the data, encrypts it according to the access policies, and uploads it to the Blockchain. Therefore, only the users whose attributes satisfy the access policies can access the data. Moreover, shared users can request access to the data from the data owner. The data owner validates the request, generates the re-encryption key and sends it to the proxy server. The authorized proxy server can perform a re-encryption operation on the original ciphertext and upload the re-encrypted ciphertext to the Blockchain. Shared users can access the re-encrypted ciphertext using their own private keys. The shared user can verify the correctness of the ciphertext after decryption. To describe our proposed scheme clearer, we give the description of used notations in Table 1.

Table 1. Symbols and descriptions

Notation	Description
U, S	Attribute set
$AS = (\mathbb{A}, \rho)$	Access structure
$params$	System public parameters
mk	Master key, kept by the KGC
sk_S	Private key for S
m	Plaintext
CT, CT'	Ciphertext
$rk_{AS \to AS'}$	Re-encryption key from AS to AS'

5.1 Concrete Structure

The attribute-based proxy re-encryption scheme extends the attribute-based encryption scheme to provide proxy re-encryption capability while protecting the user's key and plaintext, consisting of eight algorithms, including

$$(SETUP, KEYGEN, ENC, DEC, RKGEN, REENC, DEC_{re}, CLAIM).$$

$SETUP(1^k, U) \to (params, mk)$: On input of security parameters and an attribute set, the KGC generates system public parameters and Keeps the master key secret.

- Generate a tuple (e, G_1, G_2, g, p), where G_1, G_2 are two finite cyclic group of order p and g is the generator of G_1. The message space is denoted as $\{0,1\}^k$
- Choose $\alpha, a \in Z_p^*, f_1, f_2, ..., f_U, u, v, h \in G_1$ randomly.
- Select two hash function $H_1 : G_2 \to \{0,1\}^{2k}, H_2 : \{0,1\}^* \to Z_p^*$.
- Compute $mk = g^\alpha$.
- Publish system public parameters.

$$params = \{e, G_1, G_2, g, f_1, f_2, ..., f_U, u, v, h, g^a, e(g,g)^\alpha, H_1, H_2\}$$

$KEYGEN(S, mk) \to (sk_S)$: Once receiving a user attribute set and the master key, the KGC generates the user's private key.

- Choose $s \in Z_P^*$ randomly.
- Compute $sk_S = (S, K_1 = g^\alpha g^{as}, K_2 = g^s, \{K_x = f_x^s\}_{\forall x \in S})$.

$ENC(AS, m) \to (CT)$: Once receiving an access structure and a plaintext, the data owner generates the corresponding ciphertext.

- Choose $R \in \{0,1\}^k$ randomly and computes $r = H_2(m||R)$.

- Random select a vector $\vec{\mu} = (r, y_2, ..., y_n) \in Z_p^n$, where $y_2, ..., y_n \in Z_p$. For each row $\mathbb{A}_j$ of $\mathbb{A}$, computes $\lambda_j = \vec{\mu} \cdot \mathbb{A}_j, j \in [1, l]$.
- Randomly chooses $r_j \in Z_p$ for each $j \in [1, l]$. Computes

$$C = (m||R) \oplus H_1(e(g, g)^{\alpha r}), C_1 = g^r, C_2 = h^r,$$

$$\{C_{3,j} = g^{a\lambda_j} f_{\rho(j)}^{-r_j}, C_{4,j} = g^{r_j}\}_{\forall j \in [1, l]}, \tag{4}$$

$$\bar{C} = u^{H_2(m)} v^{H_2(R)}$$

- Output $CT = \{(\mathbb{A}, \rho), C, C_1, C_2, \{C_{3,j}, C_{4,j}\}_{j \in [1, l]}, \bar{C}\}$

$DEC(sk_S, CT) \rightarrow (m/\perp)$: Once receiving the user's private key and ciphertext, the recipient computes the plaintext or it outputs reject otherwise.

- If $R(S, AS) = 0$, output $\perp$,
- Otherwise, let $J = \{j : \rho(j) \in S\}$, finds elements $\theta_j \in Z_p^*$ such that $\sum_{j \in J} \theta_j \cdot A_j = (1, 0, ..., 0)$.
- Compute

$$\Delta = \frac{e(K_1, C_1)}{\prod_{j \in J} (e(K_2, C_{3,j}) \cdot e(K_{\rho(j)}, C_{4,j}))^{\theta_j}}.$$

$$m||R = C \oplus H_1(\Delta).$$

- Outputs m if $\bar{C} = u^{H_2(m)} v^{H_2(R)}$. Otherwise, outputs $\perp$.

$RKGEN(sk_S, AS') \rightarrow (rk_{AS \rightarrow AS'})$: Once receiving the user's private key and an access policy, the recipient generates a re-encryption key.

- Randomly chooses $\chi \in G_2, \delta \in Z_p^*$.
- Computes

$$rk_0 = e(g, g)^{\alpha H_2(\chi)}, rk_1 = K_1^{H_2(\chi)} \cdot h^\delta,$$

$$rk_2 = g^\delta, rk_3 = K_2^{H_2(\chi)},$$

$$\{rk_{4,j} = K_x^{H_2(\chi)}\}_{\forall x \in S}, \tag{5}$$

$$rk_5 = ENC(AS', \chi).$$

- Output the re-encryption key as $rk_{AS \rightarrow AS'} = (rk_0, rk_1, rk_2, rk_3, \{rk_{4,x}\}_{x \in S}, rk_5)$.

$REENC(rk_{AS \rightarrow AS'}, CT) \rightarrow (CT')$: Once receiving an original ciphertext and a re-encryption key, the proxy generates the re-encrypted ciphertext.

- If $R(S, AS) = 0$, output $\perp$,
- Otherwise, let $J = \{j : \rho(j) \in S\}$, finds elements $\theta_j \in Z_p^*$ such that $\sum_{j \in J} \theta_j \cdot A_j = (1, 0, ..., 0)$.

– Compute

$$C_1' = \frac{e(rk_1, C_1) \cdot e(rk_2, C_2)^{-1}}{\prod_{j \in J} (e(rk_3, C_{3,j}) \cdot e(rk_{4,\rho(j)}, C_{4,j}))^{\theta_j}}.$$

– Sets $C' = C, \bar{C}' = \bar{C}, C_2' = rk_5, C_3' = rk_0$.
– Output the re-encrypted ciphertext $CT' = (C', \bar{C}', C_1', C_2', C_3')$.

$DEC_{re}(sk_{S'}, CT') \rightarrow (m)$: Once receiving a private key and a re-encrypted ciphertext, the shared user computes the plaintext or it outputs reject otherwise.

– First decrypt χ from C_2' by running $DEC(sk_{S'}, C_2')$.
– Then computes

$$m||R = C' \oplus H_1(C_1'^{1/H_2(\chi)}).$$

$CLAIM(CT, rk_{AS \rightarrow AS'}, CT') \rightarrow (True/False)$: Once receiving an original ciphertext, re-encryption key, re-encrypted ciphertext, and proof published by a shared user, the verifier announces whether it is correct or not.

– Suppose the re-encrypted ciphertext is claimed to be wrong, where the proof is $\pi = (m, R)$.
– Check whether

$$\bar{C}' = \bar{C} = u^{H_2(m)} v^{H_2(R)}$$

$$C_1' = C_3'^r$$

– If the above equations hold, outputs *false*. Otherwise, outputs *true*.

5.2   Correctness of the Proposal Scheme

In this section, we prove the correctness of algorithms DEC and DEC_{re}, which support our proposed scheme in basis.

on input sk and CT, when $R(S, AS) = 1$, we have

$$
\begin{aligned}
\Delta &= \frac{e(K_1, C_1)}{\prod_{j \in J} (e(K_2, C_{3,j}) \cdot (K_{\rho(j)}, C_{4,j}))^{\theta_j}} \\
&= \frac{e(g^\alpha g^{as}, g^r)}{\prod_{j \in J} (e(g^s, g^{a\lambda_j} f_{\rho(j)}^{-r_j}) \cdot (f_{\rho(j)}^s, g^{r_j}))^{\theta_j}} \\
&= \frac{e(g, g)^{\alpha r} \cdot e(g^{as}, g^r)}{e(g^s, g^a)^{\sum_{j \in J} \lambda_j \theta_j}} \\
&= e(g, g)^{\alpha r}
\end{aligned}
\tag{6}
$$

$$
\begin{aligned}
C \oplus H_1(\Delta) &= [(m||R) \oplus H_1(e(g,g)^{\alpha r})] \oplus H_1(e(g,g)^{\alpha r}) \\
&= m||R
\end{aligned}
\tag{7}
$$

On input sk and CT', we have

$$
\begin{aligned}
C_1' &= \frac{e(rk_1, C_1) \cdot e(rk_2, C_2)^{-1}}{\prod_{j \in J} (e(rk_3, C_{3,j}) \cdot e(rk_{4,\rho(j)}, C_{4,j}))^{\theta_j}} \\
&= \frac{e(K_1^{H_2(\chi)} \cdot h^\delta, C_1) \cdot e(g^\delta, h^r)}{\prod_{j \in J} (e(K_2^{H_2(\chi)}, C_{3,j}) \cdot (K_{\rho(j)}^{H_2(\chi)}, C_{4,j}))^{\theta_j}} \\
&= \frac{e(K_1^{H_2(\chi)}, C_1)}{\prod_{j \in J} (e(K_2^{H_2(\chi)}, C_{3,j}) \cdot (K_{\rho(j)}^{H_2(\chi)}, C_{4,j}))^{\theta_j}} \\
&= e(g,g)^{\alpha r H_2(\chi)}
\end{aligned}
\tag{8}
$$

$$
\begin{aligned}
C' \oplus H_1(C_1'^{1/H_2(\chi)}) &= [(m \| R) \oplus H_1(e(g,g)^{\alpha r})] \oplus H_1(e(g,g)^{\alpha r H_2(\chi)/H_2(\chi)}) \\
&= m \| R
\end{aligned}
\tag{9}
$$

6   Security Analysis

In this section, we analyze the security of the proposed scheme. Specifically, following the security requirements discussed earlier, our analysis focuses on Selective-Structure Chosen Plaintext Security, Key Security, and Verifiability for the proposed scheme.

6.1   Selective-Structure Chosen Plaintext Security

Theorem 1. *Our scheme satisfies semantically secure assuming that the q-parallel BDHE assumption holds.*

Proof. During the data uploading phase, the confidentiality of the data owner is achieved based on our encryption algorithm, which is modified from a ciphertext-policy attribute-based encryption scheme. Assume there exists a PPT adversary $\mathcal{A}$ who can break the Selective-Structure Chosen Plaintext Security in our scheme with a non-negligible probability ϵ, then a simulator $\mathcal{B}$ can be constructed that can solve the q-parallel BDHE assumption with a non-negligible advantage. Recall that $\mathcal{B}$ is given a q-parallel BDHE instance $(\vec{v}, T)$, and its goal is to decide whether T equals to $e(g,g)^{r \cdot \alpha^{q+1}}$ or T is randomly chosen from G_2.

In the data-sharing phase, the original data recipient generates a re-encryption key to send to the proxy server. In this case, the proxy key consists of two parts: a valid key that has been blinded, and a blinding factor that is encrypted under the shared user access policy. The blinding factor has the same privacy protection as plaintext messages, and its security is provided by the CP-ABE scheme. The proxy decrypts the original ciphertext using the blinded key and computes the decryption factor in the blinded state. Nevertheless, the proxy still cannot get any information about the plaintext.

6.2   Key Security

Collusion resistance refers to the property of a system or protocol that prevents participants from collaborating in a way that undermines the intended security or fairness guarantees. It ensures that even if the proxy and shard user conspires together, they cannot manipulate the system to gain unfair advantages or obtains the original data recipient's private key. In our proxy re-encryption scheme, the re-encryption key is masked by h^δ. If the proxy colludes with the shared user, they can compute the blinding factor χ to un-blind the user's private key. However, thanks to the protection of h^δ, the adversary still cannot obtain the original valid key. Therefore, our scheme can resist collusion attacks and achieves key security.

6.3   Verifiability

Theorem 2. *The proposed scheme satisfies the verifiable property assuming the discrete logarithm assumption holds.*

Proof. In the scheme proposed in this paper, the ciphertext and the re-encrypted ciphertext are stored on the Blockchain. Using the tamper-proof nature of the Blockchain, the data can be guaranteed not to be tampered with. Meanwhile, the proxy re-encryption scheme used in this paper adds a digital commitment to the message content in the encryption algorithm as well as an authentication mechanism, which makes the verifier believe that the correct re-encryption algorithm computes the ciphertext and that the proxy cannot pass the authentication by using other ciphertexts. Suppose a PPT adversary $\mathcal{A}$ can violate verifiability of our proposed scheme with a non-negligible advantage ϵ, then a simulator $\mathcal{B}$ can be constructed that can solve the discrete logarithm problem with a non-negligible advantage. Recall that $\mathcal{B}$ is given a discrete logarithm instance $(e, G_1, G_2, p, g, g^\beta)$, its goal is to output the value β.

7   Performance Analysis

In this section, we present the computation costs and communication overheads of the proposed scheme, and compare it with some related schemes, such as [7] and [6].

7.1   Functionality Comparison

The functionality comparison between our scheme and other related works is shown in Table 2. "✓" indicates that the feature can be implemented while "✗" signifies that the scheme fails to realize this functionality.

Fine-grained access control indicates that a data owner can specify a class of recipients to access the data whose attributes conform to a certain access policy. That is one-to-many access control. And the scheme of [7] belongs to identity-based encryption, which is suitable for one-to-one encryption. All of the

Table 2. Functionality comparison of our scheme and related works.

Functionality	[7]	[6]	Our scheme
Fine-Grained Access Control	✗	✓	✓
Data Sharing	✓	✓	✓
Verifiability	✗	✓	✓
Collusion-Resistance	✓	✓	✓
Non-repudiation	✓	✓	✓

above schemes combine proxy re-encryption techniques to realize data sharing, where the original data recipient generates the re-encryption key and undergoes a re-encryption operation by an incompletely trustworthy third-party proxy server, which realizes the conversion of ciphertexts without exposing the plaintext information. Verifiability prevents lazy proxies from dishonestly performing re-encryption algorithms to save costs while randomly selecting messages from the ciphertext space to ensure the interests of shared users. There is no effective authentication mechanism provided in [7]. The user can only repeat the re-encryption algorithm to compare the ciphertexts. The collusion-resistance property secures the key of the original data recipient and prevents curious proxy and shared users from colluding to compute the complete user's private key. Non-repudiation guarantees users' rights and interests through the tamper-proof nature of Blockchain technology, providing a credible source of information when users have disputes with proxy.

7.2 Theoretical Analysis

We theoretically analyze and compare the computation and storage cost of our works and other related works in Table 4 and Table 3. We mainly focus on the most time-consuming calculations in bilinear groups, including exponentiation operation, multiplication operation, and bilinear pairings. We set $\mathcal{P}$, $\mathcal{E}$, and $\mathcal{M}$ to represent the computation cost of a bilinear pairing, a modular exponentiation, and a modular multiplication operation, respectively. The size of a single group element in G_1 and G_2 are correspondingly denoted as $|G_1|$ and $|G_2|$.

Table 3. Storage costs for messages in each phase of the protocol.

Scheme	private key	ciphertext	re-key	re-encrypted ciphertext
[7]	$|G_1|$	$|G_1| + |G_2|$	$2|G_1| + |G_2|$	$2|G_1| + 2|G_2|$
[6]	$(3n+2)|G_1|$	$(4n+2)|G_1| + |G_2|$	$(2n+5)|G_1|$	$(n+4)|G_1| + |G_2|$
Our scheme	$(n+2)|G_1|$	$(2n+3)|G_1| + |G_2|$	$(3n+4)|G_1| + 2|G_2|$	$(2n+2)|G_1| + 3|G_2|$

Storage Costs. There are four types of messages in our scheme. During the registration phase, KGC generates a user's secret key to send to the user with

Table 4. Computation costs for algorithms in each phase of the protocol.

Scheme	KEYGEN	ENC	DEC	RKGEN	REENC	DEC_{re}	CLAIM
[7]	$\mathcal{E}$	$\mathcal{P}+2\mathcal{E}+\mathcal{M}$	$\mathcal{P}+\mathcal{E}$	$\mathcal{P}+2\mathcal{E}+2\mathcal{M}$	$\mathcal{P}+\mathcal{M}$	$2\mathcal{P}+2\mathcal{M}$	$\mathcal{P}+\mathcal{M}$
[6]	$(4n+2)\mathcal{E}+(2n+1)\mathcal{M}$	$(5n+3)\mathcal{E}+(n+1)\mathcal{M}$	$(2n+1)\mathcal{P}+n\mathcal{E}+2n\mathcal{M}$	$6\mathcal{E}+3\mathcal{M}$	$(3n+1)\mathcal{P}+n\mathcal{E}+(3n+2)\mathcal{M}$	$3\mathcal{P}+3\mathcal{M}$	$n\mathcal{P}+n\mathcal{E}$
Our scheme	$(n+2)\mathcal{E}+\mathcal{M}$	$(3n+5)\mathcal{E}+(2n+1)\mathcal{M}$	$(2n+1)\mathcal{P}+n\mathcal{E}+2n\mathcal{M}$	$(4n+8)\mathcal{E}+(n+1)\mathcal{M}$	$(2n+2)\mathcal{P}+n\mathcal{E}+(2n+1)\mathcal{M}$	$(2n+1)\mathcal{P}+(n+1)\mathcal{E}+2n\mathcal{M}$	$3\mathcal{E}+\mathcal{M}$

a key size of $(n+2)|G_1|$. The data owner protects the data using an encryption algorithm that generates a ciphertext of size $(2n+3)|G_1|+|G_2|$. When data is shared, the original data recipient first generates a re-encryption key with size $(3n+4)|G_1|+2|G_2|$. Then the agent performs the re-encryption operation to generate the re-encryption ciphertext of size $(2n+2)|G_1|+3|G_2|$. [6] and our scheme are attribute-based encryption schemes where the user's secret key and ciphertext are associated with attributes as well as access policies, respectively. The message length increases linearly with the number of attributes, enabling fine-grained one-to-many access control. While [7] provides identity-based encryption, the complexity of its storage overhead will be the same as ours when the identity is converted to user attributes. Since [7] belongs to identity-based encryption and [6] re-encrypts attribute-based ciphertext agents into identity-based cryptosystems, their re-encryption key has a slight advantage over our scheme, which approximates the size of the re-encryption key in our scheme to be twice as large as theirs. However, they add extra parameters to the user's private key and ciphertext in order to provide verifiability, resulting in all other message lengths being larger than ours. Therefore, the scheme proposed in this paper has a low storage overhead when implementing fine-grained access control and data sharing.

Computation Costs. In the proposed scheme, KGC performs the $KEYGEN$ algorithm to compute the user's private key, which costs $(n+2)Te+Tm$. During the data uploading phase, the data owner encrypts its data by executing the ENC algorithm, which costs $(3n+5)Te+(2n+1)Tm$. Original data recipient access data by executing the DEC algorithm, which costs $(2n+1)Tp+nTe+2nTm$. DR generates a re-encryption key by executing the $RKGEN$ algorithm, which costs $(4n+8)Te+(n+1)Tm$. And proxy executing the $REENC$ algorithm to compute a re-encrypted ciphertext, which costs $(2n+2)Tp+nTe+(2n+1)Tm$. Then, SU requests and decrypts ciphertext by executing the DEC_{re} algorithm, which costs $(2n+1)Tp+(n+1)Te+2nTm$. The $CLAIM$ algorithm is run when a shared user verifies a re-encrypted ciphertext, which costs $3Te+Tm$. Comparing the attribute-based scheme [6] and our scheme, the encryption and decryption algorithms have the same computational complexity, but in the verification phase of re-encrypted ciphertexts, our proposed scheme is efficient.

7.3 Experimental Analysis

In order to evaluate the execution efficiency of the scheme proposed in this paper, we design experiments to test the time overhead required for each phase of the scheme. We conducted the experiment in Java running on the PC with one 2.30 GHz Intel Core i5, 16.0-GB memory, and Windows 11 system. We use an elliptic curve of type A and set the security parameter to 160bit.

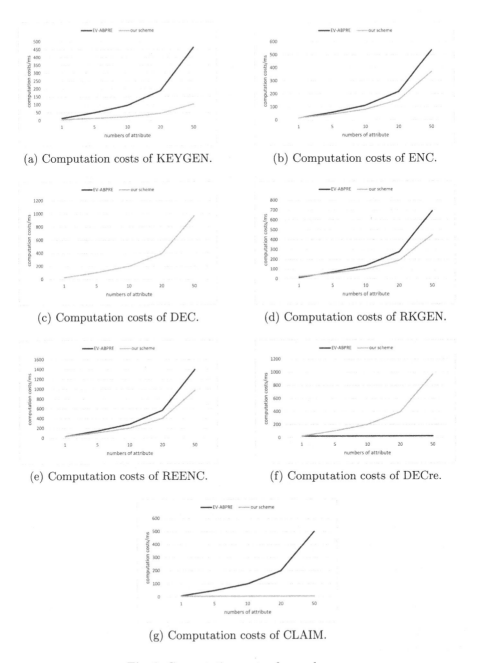

(a) Computation costs of KEYGEN.

(b) Computation costs of ENC.

(c) Computation costs of DEC.

(d) Computation costs of RKGEN.

(e) Computation costs of REENC.

(f) Computation costs of DECre.

(g) Computation costs of CLAIM.

Fig. 2. Computation costs of our schemes.

Figure 2 compares the execution time of our proposed scheme against a related scheme [6].

Figure (2a, 2b, 2d, 2e) shows that our scheme is more efficient than the EV-ABPRE scheme in the key generation, encryption, re-encryption key generation, and re-encryption phases. This is because their scheme needs to compute additional parameters to achieve verifiability. It is shown in Fig. 2c that both schemes have the same decryption efficiency, which is equivalent to performing a traditional CP-ABE decryption. Figure 2f shows that our scheme has a larger computational overhead in performing the decryption of the re-encrypted ciphertext because in [6], the re-encrypted ciphertext is converted to a single identity-based encryption, whose decryption efficiency is independent of the number of attributes. Figure 2g shows that our scheme has a significant advantage in the ciphertext verification phase, and the computational overhead does not grow linearly with the number of attributes increases.

8   Conclusion

This paper applies Blockchain and attribute-based proxy re-encryption to propose a privacy-preserving security model in connected vehicle data sharing. ABPRE allows third-party cloud servers to transform ciphertexts without obtaining plaintext information to achieve user access control to data. At the same time, to solve the problem that centralized cloud servers may have a single point of failure and dishonest cloud servers return wrong ciphertexts to save costs, this paper combines Blockchain technology to provide tamper-proof data storage and a verifiable mechanism of re-encrypted ciphertexts to protect users' interests from being infringed. Security and performance analyses demonstrate that the proposed scheme achieves a good balance between security and efficiency.

References

1. Ateniese, G., Fu, K., Green, M., Hohenberger, S.: Improved proxy re-encryption schemes with applications to secure distributed storage. ACM Trans. Inform. Syst. Secur. (TISSEC) **9**(1), 1–30 (2006)
2. Baker, T., Asim, M., Samwini, H., Shamim, N., Alani, M.M., Buyya, R.: A blockchain-based fog-oriented lightweight framework for smart public vehicular transportation systems. Comput. Netw. **203**, 108676 (2022). https://doi.org/10. 1016/j.comnet.2021.108676
3. Bao, Y., Qiu, W., Cheng, X., Sun, J.: Fine-grained data sharing with enhanced privacy protection and dynamic users group service for the IoV. IEEE Trans. Intell. Transp. Syst. 1–15 (2022). https://doi.org/10.1109/TITS.2022.3187980
4. Blaze, Matt, Bleumer, Gerrit, Strauss, Martin: Divertible protocols and atomic proxy cryptography. In: Nyberg, Kaisa (ed.) EUROCRYPT 1998. LNCS, vol. 1403, pp. 127–144. Springer, Heidelberg (1998). https://doi.org/10.1007/BFb0054122
5. Elagin, V., Spirkina, A., Buinevich, M., Vladyko, A.: Technological aspects of blockchain application for vehicle-to-network. Information **11**(10), 465 (2020)

6. Feng, T., Wang, D., Gong, R.: A blockchain-based efficient and verifiable attribute-based proxy re-encryption cloud sharing scheme. Information **14**(5) (2023). https://doi.org/10.3390/info14050281

7. Gao, Y., Chen, Y., Hu, X., Lin, H., Liu, Y., Nie, L.: Blockchain based IIoT data sharing framework for SDN-enabled pervasive edge computing. IEEE Trans. Industr. Inf. **17**(7), 5041–5049 (2021). https://doi.org/10.1109/TII.2020.3012508

8. Ge, C., Liu, Z., Xia, J., Fang, L.: Revocable identity-based broadcast proxy re-encryption for data sharing in clouds. IEEE Trans. Dependable Secure Comput. **18**(3), 1214–1226 (2021). https://doi.org/10.1109/TDSC.2019.2899300

9. Ge, C., Susilo, W., Baek, J., Liu, Z., Xia, J., Fang, L.: A verifiable and fair attribute-based proxy re-encryption scheme for data sharing in clouds. IEEE Trans. Dependable Secure Comput. **19**(5), 2907–2919 (2021)

10. Green, Matthew, Ateniese, Giuseppe: Identity-based proxy re-encryption. In: Katz, Jonathan, Yung, Moti (eds.) ACNS 2007. LNCS, vol. 4521, pp. 288–306. Springer, Heidelberg (2007). https://doi.org/10.1007/978-3-540-72738-5_19

11. Hanaoka, G., et al.: Generic construction of chosen ciphertext secure proxy re-encryption. In: Dunkelman, Orr (ed.) CT-RSA 2012. LNCS, vol. 7178, pp. 349–364. Springer, Heidelberg (2012). https://doi.org/10.1007/978-3-642-27954-6_22

12. Karthikeyan, H., Usha, G.: Real-time DDoS flooding attack detection in intelligent transportation systems. Comput. Electr. Eng. **101**, 107995 (2022). https://doi.org/10.1016/j.compeleceng.2022.107995

13. Kenney, J.B.: Dedicated short-range communications (DSRC) standards in the United States. Proc. IEEE **99**(7), 1162–1182 (2011)

14. Liang, X., Cao, Z., Lin, H., Shao, J.: Attribute based proxy re-encryption with delegating capabilities. In: Proceedings of the 4th International Symposium on Information, Computer, and Communications security, pp. 276–286 (2009)

15. Liu, J., Zhang, L., Li, C., Bai, J., Lv, H., Lv, Z.: Blockchain-based secure communication of intelligent transportation digital twins system. IEEE Trans. Intell. Transp. Syst. **23**(11), 22630–22640 (2022). https://doi.org/10.1109/TITS.2022.3183379

16. Nakamoto, S.: Bitcoin: a peer-to-peer electronic cash system. Decentralized business review (2008)

17. Ohata, S., Kawai, Y., Matsuda, T., Hanaoka, G., Matsuura, K.: Re-encryption verifiability: how to detect malicious activities of a proxy in proxy re-encryption. In: Nyberg, K. (ed.) Topics in Cryptology – CT-RSA 2015, pp. 410–428. Springer, Cham (2015)

18. Panigrahy, S.K., Emany, H.: A survey and tutorial on network optimization for intelligent transport system using the internet of vehicles. Sensors **23**(1), 555 (2023)

19. Shao, J.: Anonymous id-based proxy re-encryption, pp. 364–375 (2012). https://doi.org/10.1007/978-3-642-31448-3_27

20. Xu, L., Wu, X., Zhang, X.: CL-PRE: a certificateless proxy re-encryption scheme for secure data sharing with public cloud. In: Proceedings of the 7th ACM Symposium on Information, Computer and Communications Security. pp. 87–88. ASIACCS 2012. Association for Computing Machinery, New York (2012). https://doi.org/10.1145/2414456.2414507

21. Zavvos, E., Gerding, E.H., Yazdanpanah, V., Maple, C., Stein, S., Schraefel, M.: Privacy and trust in the internet of vehicles. IEEE Trans. Intell. Transp. Syst. **23**(8), 10126–10141 (2022). https://doi.org/10.1109/TITS.2021.3121125

Blockchain-Based Hierarchical Access Control with Efficient Revocation in mHealth System

Ting Liang[1], Yaorui He[1], Pei Huang[1], and Zhe Xia[1,2(✉)]

[1] School of Computer Science and Artificial Intelligence, Wuhan University of Technology, Wuhan, China

[2] Hubei Key Laboratory of Transportation Internet of Things, Wuhan University of Technology, Wuhan 430071, China
xiazhe@whut.edu.cn

Abstract. With the development of information technology, people can share their health records (PHRs) through the Internet and obtain rapid medical services, which makes mobile health become a promising field. PHRs are collected from wireless body area networks (WBANs) and will be shared with people in different fields through public channels, increasing the risk of leaking personal privacy. Ciphertext-policy attribute-based encryption (CP-ABE) is a popular solution for fine-grained access control, but most existing schemes cannot be directly applied to the WBANs with limited resources and dynamic changes in user roles. In this paper, to meet the requirement of the mHealth System, we propose blockchain-based hierarchical access control with efficient revocation in the mHealth system. We use the Extendable Hierarchical attribute-based encryption (EH-ABE), a file-related hierarchical access control scheme, to encrypt PHRs, which reduces the repetitive computation and storage overhead. The proposed scheme adds the function of offline/online encryption, which can greatly save the energy consumption of the sensors in the WBANs. In addition, this scheme can provide attribute-level user revocation and is proven to be IND-CCA secure.

Keywords: Attribute level user revocation · Hierarchical Access Control · Offline/Online Encryption

1 Introduction

With the development of information technologies, wireless body area networks (WBANs) have been used to provide higher quality medical services [15]. WBAN is the fundamental component in mobile health system consisting of smart medical sensors. It can collect physiological parameters, such as body temperature, blood glucose level, blood pressure, heart rate, Electrocardiogram (ECG), or Electromyogram (EMG) [9]. These body parameters can be used to monitor

J. Chen and Z. Xia (Eds.), BlockTEA 2023, LNICST 577, pp. 135–153, 2024.
https://doi.org/10.1007/978-3-031-60037-1_8

chronic diseases. The mobile device receives the personal health records (PHRs) from the body area networks and encrypts them before uploading them to the cloud service provider (CSP). The PHR user, such as a researcher in a medical college, or a doctor in the hospital, can access these data if he/she is a valid user [17]. Then, the data could be used for better medical research or to provide users with more effective treatments.

PHR involves the user's private information and is shared with people in different fields through public channels. There may be malicious users in the system who steal data to obtain economic benefits, but the leakage of PHR can pose a serious threat to the life and health of patients. Therefore, ensuring data confidentiality is particularly crucial [22]. According to the multi-receiver characteristic of PHR sharing, CP-ABE is the best solution because it can provide fine-grained access control. However, the original CP-ABE schemes [2,28], are not suitable for direct application in the mHealth system. So far many CP-ABE schemes [13,23,26,29,32] have been proposed for the mHealth system. But none of them offer the function of user revocation. There are also some schemes with user revocation [1,8,24,30,31] which only provide user revocation but not user's attribute revocation. The work on attribute-level user revocation [4] is relatively less.

In the existing work, access policies are isolated and fixed during encryption, resulting in a complicated and repetitive encryption process, which is not suitable for resource-constrained body area networks. For example, a patient may have high blood pressure and heart disease. To treat these chronic diseases, physical signs information needs to be monitored, such as body temperature, blood pressure, heart rate, Electrocardiogram (ECG), and Electromyogram (EMG). Different data may require different access policies, such as Fig. 1. Body temperature can be accessed by physicians and nurses in the Department of Cardiology and Orthopedics. T_2 is an access policy for blood pressure, and doctors who treat hypertension and heart disease need to be allowed to access these data. T_4 is the access strategy for heart rate, and T_5 is the access strategy for EMG. However, hierarchical relationships may exist between these policies. For instance, doctors treating high blood pressure also need to monitor the patient's blood pressure and temperature. Doctors who treat cervical spondylosis need to monitor their patients' EMG and heart rate. The cardiologist needs blood pressure, temperature, ECG, EMG, and heart rate. In traditional schemes, they are encrypted separately, and users need to have an access policy for each data provider that provides services to them. In the hierarchical access control scheme, some access policies with hierarchical relationships can be aggregated into one, and data is simultaneously encrypted under the same access policy. This method is helpful to improve the efficiency of encryption and reduce computation and storage costs.

In this paper, we propose a revocable CP-ABE scheme with hierarchical access control and offline/online encryption, named RH-ABE, which simultaneously solves the problems of encryption efficiency, repetitive computation, and attribute-level user revocation. Inspired by the Key Encrypting Key(KEK) technique in the efficient revocation [11], we propose a revocable hierarchical access

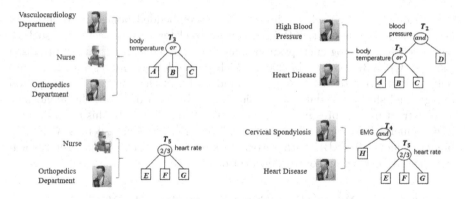

Fig. 1. Access Policies

control scheme for the mHealth system based on EH-ABE in [28]. RH-ABE enables the PHR owner(PO) to precompute policy-related ciphertext under self-defined fine-grained access policies when the body area network device is offline. When receiving PHR from sensors, the PO performs the online encryption process to compute the ciphertext associated with the message before uploading it to the cloud service provider (CSP). Unauthorized users cannot get the plaintext PHRs. By the revocation method, the PHR is preserved from being abused and exposed to revoked users. Our main contributions can be summarized as follows.

- **Offline/Online encryption:** Various sensor devices in body area networks have limited storage capacity, small battery capacity, and low computing power. In our scheme, when the device is offline, the PO encrypts the ciphertext related to the policy in advance. When the device is online, it encrypts the received ciphertext related to the PHR. This approach is helpful to reduce the sensor devices' energy consumption, hence it is suitable to be used in the resource-constrained environment.
- **Hierarchical access control:** In our scheme, we use a hierarchical access control structure that can encrypt a series of files with hierarchical access relationships. It can reduce the cost of complicated and repetitive encryption.
- **Attribute-level user revocation:** In this paper, we can implement fine-grained user revocation, where some attributes of a user are revoked without affecting the permissions of other attributes. This is suitable for a situation where the user's role is changing dynamically, with some attributes being removed while others remain, such as the change of the doctor's position. Once the attribute of a user is revoked, if his attributes do not match the access policy, he cannot continue to access the previous ciphertext.

2 Related Works

Attribute-based encryption (ABE) was first introduced by Sahai and Waters [21] in 2003. Then, Goyal et al. [6] proposed the first key-policy attribute-based encryption (KP-ABE) scheme in 2006, which embedded policies into keys and attributes into ciphertext. KP-ABE can be used for various scenarios, such as video websites and log encryption management. Bethencourt et al. [2] proposed the first CP-ABE scheme in 2007, which embedded attributes into keys and policies into ciphertext. CP-ABE is found useful in the encrypted storage and fine-grained sharing of data on public clouds. In both of them, the monotonic access structure is expressed using a tree structure. Thanks to this feature, some improvements have been proposed for different requirements, such as ABE with hidden policy [19], ABE with revocable storage [12], ABE with user revocation [18], ABE with outsourced decryption [7].

2.1   Attribute-Based Hierarchical Encryption (HABE)

Attribute-based hierarchical control schemes can be divided into two classes according to the different construction ideas. One is the hierarchical management of attributes to reduce the workload of central attribute authority. The other is to aggregate and encrypt files with hierarchical access relationships. There are many schemes [14,20,25] followed the first idea by setting a central authority to govern a series of hierarchical domain authorities and the key generation could be performed by all of the authorities which can avoid key abuse. In 2016, Wang et al. [27] first proposed a hierarchical attribute-based encryption scheme in which encrypted files are hierarchical access control relationships. This kind of HABE can reduce the burden on data owners by encrypting files with hierarchical access control relationships. Following this idea, the access policies associated with the hierarchical files can be integrated into a single access structure with several levels. It is suitable for the mHealth system which has High requirements for resources and efficiency. In the existing work, there are several efficient schemes are proposed [5,28]. But the [5] is not flexible and only considers an AND-gate access structure.

2.2   User Revocation

A series of fine-grained access control constructions with user revocation for the mHealth system was proposed.

Based on the different ways to implement the revocation, user revocation can be done either by direct revocation or by indirect revocation. The first approach is achieved by maintaining a list of revocation users. Yang et al. [30] used this way by embedding the identity in the secret key and storing the revocation list in the public cloud. Tan et al. [24] proposed a blockchain-empowered approach for data sharing. In their scheme, the malicious user's ID will be added to the revocation list, and the revocation list will be sent to the medical practitioner. The medical practitioner re-encrypts the data. Li et al. [17] proposed a traceable and revocable scheme in which user revocation is complicated. The valid user in their scheme needs to be pre-chosen, but it's not suitable for most situations the PO doesn't know the users who can match the access policy. The second approach is realized by updating the secret key. Only the updated user can decrypt successfully, and the revoked user cannot be updated. Guo et al. [8] proposed a revocable scheme in the cloud-assisted IoMT system. In the revocation process, a TTP updates the attribute set for the revoked user. The revoked user cannot get the information for a valid secret key.

Based on the scope of influence, attribute revocation can be divided into three categories: user revocation, user partial attribute revocation, and system attribute revocation. They are described as follows:

- **User revocation**: All attributes of a specific user are revoked without affecting other users.
- **Partial attributes revocation of a user**: Revocation of the attributes of a specific user without affecting other users and other attributes of the user.
- **System attribute revocation**: A specific attribute is revoked, and all users no longer have permission on the attribute.

Most of the existing works are user revocation, which is not the best solution for the changeable environment. In the scheme [16], there are two kinds of users: public domain users and individual domain users. The revocation of a public domain user can achieve attribute revocation and user revocation. The scheme [4] is also an attribute-level revocable scheme that used the Key Encrypting Key(KEK) techniques proposed in [11], but the access structure is the ordered binary decision diagram (OBBD).

3 Preliminaries

In this section, we review some basic notions and definitions. They are the basic knowledge of cryptography and effective tools for building the scheme.

3.1 Bilinear Maps

Definition 1 (Bilinear Maps [3]). Let G_0 and G_T be two finite cyclic groups of prime order p. And g is denoted as a generator of G_0. Select a map $e : G_0 \times G_0 \rightarrow G_T$ should satisfy:

- Bilinear: $\forall P, Q \in G_0$ and $\forall a, b \in Z_p^*$, there is $e(P^a, Q^b) = e(P, Q)^{ab}$,
- Non-degenerate: $e(g, g) \not\equiv 1$,
- Computable: $\forall P, Q \in G_0$, $e(P, Q)$ can be computed efficiently.

3.2 Access Structure

Definition 2 (Access Structure [28]). Denote $\{P_1, P_2, \cdots P_n\}$ as a set of parties. A collection $\mathbb{A} \subseteq 2^{\{P_1, P_2, \cdots P_n\}}$ is monotone if that $\forall B, C, B \in \mathbb{A}$ and $B \subseteq C$ then $C \in \mathbb{A}$. An access structure is a collection $\mathbb{A}$ of non-empty subsets of $\{P_1, P_2, \cdots P_n\}$ i.e., $\mathbb{A} \subseteq 2^{\{P_1, P_2, \cdots P_n\}} \setminus \{\emptyset\}$. The sets in $\mathbb{A}$ are called the authorized sets, and the sets not in $\mathbb{A}$ are called the unauthorized sets.

3.3 Hierarchical Access Tree

Definition 3 (Hierarchical Access Tree [28]**).** The hierarchical access tree defined in [6] is used that is an extension of a tree structure in [28]. Denote $\mathcal{T}$ as a hierarchical access tree with an access structure that can be further represented as a series of hierarchical access structures. The x represents the non-leaf node of the tree, and y represents the leaf node of the tree. Each sub-tree with root x in $\mathcal{T}$ is a sub-structure. For example in Fig. 2, in the integrated structure T_1, there are four sub-trees that the trees T_2, T_3, T_4 and T_5 rooted at node 2, 3, 4 and 5 represent different sub-structures. Each node x of $\mathcal{T}$ that is not a leaf stands for a threshold gate i.e. an AND, OR, or OUTOF-gate, according to the number of its children and a threshold value. Denote num_x as the number of children of x, and z_x as the threshold value, then we have $0 < z_x \leq num_x$. In particular, the threshold gate is OR when $z_x = 1$, and the threshold gate is AND when $z_x = num_x$.

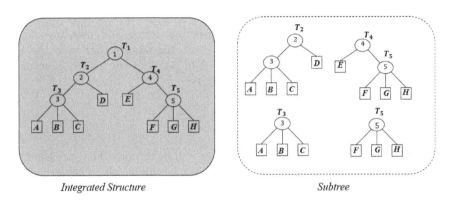

Integrated Structure   Subtree

Fig. 2. Hierarchical Access Tree

4 Models and Definitions

In this section, we outline the models and definitions, including the system model, the communication model, the adversary model, the security requirements, and the security assumptions.

4.1 System Model

As shown in Fig. 3, RH-ABE consists of five types of participants: trusted authority (TA), PHR owner (PO), mHealth cloud service provider (mHealth CSP), PHR user (PU), and blockchain. The characteristics and functions of each party are described below:

- **TA**: TA is a trusted authority that is responsible for managing users and distributing secret keys. Specifically, it produces the system's public parameters and the system master secret key. It also issues the keys for PUs which grants fine-grained access to them. The keys are associated with their respective attributes, and none of them can collude with others for invalid entry. To effectively manage users and implement attribute-level user revocation, TA needs to generate a list of valid users for each attribute. When a user needs to be removed from the system, or the user's attribute set changes, the valid user list of attributes needs to be updated immediately.
- **PO**: PO integrates the PHRs that are collected by WBSN. WBSN employs smart wireless sensors embedded inside or on the skin of a patient. These sensors monitor the patient's vital physiological parameters who is suffering from chronic diseases. The collected data is aggregated and transmitted to a mobile device via wireless communication [31]. Before uploading the PHR to the mHealth CSP, the PO should encrypt it according to the access policy.
- **mHealth CSP**: In our system, CSP has two main functions: transform and store. In this scheme, attribute-level user revocation is implemented with the help of CSP. It runs algorithms to generate KEKs for all users in the system. When receiving the data uploaded by the PO, it first performs ciphertext transformation with the attribute group key. Then, the mHealth CSP stores the transformed PHRs. It also needs to delete the Previous ciphertext. When receiving a download request from the PU, the CSP constructs a ciphertext header that binds the information of the attribute group key. If the user is not revoked, he/she can recover the correct group key by using the valid KEKs. Furthermore, the CSP needs to keep the list of unrevoked users up to date.
- **PU**: PU is a PHR user, PU can access the shared PHRs if and only if he/she is not revoked and his/her attributes match the embedded access policy.

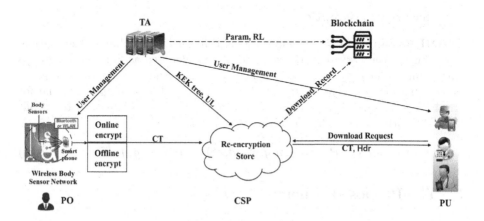

Fig. 3. System Architecture

4.2 Communication Model and Adversary Model

In our scheme, TA is assumed to send the secret keys to the users over secure channels. Information transmitted between users and the CSP is assumed to use an authentication channel. TA is fully trusted, it can't disclose any secrets. CSP is deemed as semi-honest, it keeps group keys secret and follows pre-defined operations. Meanwhile, it is curious about the PHRs and attempts to get secret information about the original PHRs as much as possible. We assume that CSP does not collude with revoked users to obtain unauthorized data. PUs are not trusted, and they can collude with each other for decrypting PHRs which none of them can decrypt alone.

4.3 Security Requirements

The following security requirements are considered in our proposed scheme.

- **Data Confidentiality.** PHRs are related to the life, health, and safety of users. The confidentiality of outsourced data should be protected. There are two main aspects of confidentiality: unauthorized users and revoked users. Unauthorized users, such as CSP or other unauthorized users, cannot access plaintext information of the outsourced PHRs. In this paper, our scheme is secure against the chosen ciphertext attacks to ensure data security.
- **Collusion Resistance.** In the system, different users' attribute sets are not identical. Some malicious users may try to collude by combining their secret keys to gain more privilege. In our scheme, this kind of attack defined as collusion can't succeed. Since TA is honest and user revocation is performed with the help of CSP. We assume that CSP can't collude with the revoked users.

4.4 Security Assumptions

CBDH Assumption [28]: Computational Bilinear Diffie-Hellman Assumption. According to the **Definition 1**, G_0 and G_T are denoted as finite cyclic groups with prime order p. Denote g as a generator of G_0. And $e : G_0 \times G_0 \to G_T$ is a bilinear pairing. a, b, c are randomly chosen from $\mathbb{Z}_p$. The assumption is that for any PPT adversary $\mathcal{A}$, with inputs $(G_0, p, g, g^a, g^b, g^c)$, it is infeasible to output $e(g, g)^{abc}$ with a non-negligible advantage, that is:

$$Adv_{\mathcal{A}} = Pr[\mathcal{A}(G_0, p, g, g^a, g^b, g^c) \to arrowe(g, g)^{abc}].\qquad(1)$$

5 The Proposed Scheme

In this section, we describe our proposed blockchain-based hierarchical access control with efficient revocation in the mHealth system. We divide the program into the following six phases: System initialization, User authorization, PHR outsourcing, Re-encryption, PHR access, and Attribute revocation. Let's explain it in detail.

System Initialization. In this phase, TA initializes the protocol and generates the public parameters. First, TA chooses the security parameter λ as input and denotes U as the attribute universe. According to the *Definition 1*, TA chooses a finite cyclic group G_0 with order p, where g and h are two generators. Denote e as a bilinear map $e : G_0 \times G_0 \rightarrow G_T$. $\Delta_{i,S} = \prod_{j \in s, j \neq i} \frac{x-j}{i-j}$ is denoted as the Lagrange coefficient. (E, D) is denoted as a symmetric cipher, and two exponents $\alpha, \beta \in \mathbb{Z}_p$ are randomly chosen. Furthermore, two cryptographic hash functions: $H_1 : \{0,1\}^* \rightarrow G_0$, $H_2 : G_T \times G_0 \times G_T \rightarrow \{0,1\}^l$ are chosen, where l is the symmetric key length. The system parameter $Param = \{G_0, G_T, e, p, g, h = g^\beta, e(g,g)^\alpha, H_1, H_2, (E, D)\}$. And the master secret key $MSK = \{\beta, g^\alpha\}$. Then, TA generates a binary KEK tree with u leaf nodes, where u is the maximum number of users in the system. For each node in the KEK tree, TA selects a random number as the KEK, we call it KEK_i. It creates a user Revocation List (RL), initialized to empty, that records the revoked users of each attribute i. It also creates a legitimate user list $UL = G_i, (i \in U)$, initialized to empty, that records the legitimate users of each attribute i. Finally, TA uploads the system parameter $Param$ and the RL to the blockchain so that all of the entities in the system can access it. The master key MSK is held by the authority secretly for user authorization. It sends the KEK tree, UL, and all of the KEK_i to the CSP over a secure channel.

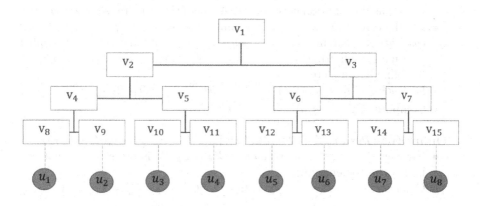

Fig. 4. hierarchical access control

User Authorization. In this phase, each user registers with TA to obtain the keys of the ABE algorithm and the KEK keys. At first, TA checks the user's attributes, adds the user to the UL, and generates the keys of the ABE algorithm as follows. Input $Param, MSK$, the set γ attributes of this user. And it chooses a random exponent: $r \in \mathbb{Z}_p$ and computes $D = g^\alpha h^r$. For each attribute $i \in \gamma$, randomly choose $r_i \in \mathbb{Z}_p$, and computes $D_i = g^r \cdot H_1(i)^{r_i}, D_i' = h^{r_i}$. The keys of ABE algorithm $SK = \{D, \{D_i, D_i'\}, \forall i \in \gamma\}$. Then, TA chooses a null leaf node i for this user and assigns all KEK keys along the path from the root node to the leaf node u_i to the user. The set of KEK keys of the path is called path keys (PK_i), such as $PK_5 = \{KEK_1, KEK_3, KEK_6, KEK_{12}\}$. As shown in Fig. 4, every user is assigned to the leaf nodes of the tree $(u_1, u_2, u_3, \cdots, u_8)$. Random keys (KEK_i) are generated and assigned to each leaf node and internal node $(v_1, v_2, v_3, \cdots, v_{15})$.

PHR Outsourcing. PHR is encrypted before uploading to CSP. Data encryption consists of two parts: offline encryption and online encryption. Before PHR is collected or WBSN is offline, PO performs the offline encryption algorithm in advance which is combined with the algorithm in [28]. When WBSN is online, the PO receives the PHRs and performs the online encryption algorithm under the chosen access policy. The process is explained in detail below.

- **OfflineEnc**$(Param, \mathcal{T}) \rightarrow (\{C_x^{1'}, C_x^2\}_{\forall x \in \mathcal{T}}, \{C_y, C_y'\}_{\forall y \in \mathcal{T}})$
  - Provided with the system parameter $Param$, the PO chooses a hierarchical access tree $\mathcal{T}$ according to the related policies. A is the root node, x is a non-leaf node and y is a leaf node.
  - It generates a polynomial q_x for each node x in the tree $\mathcal{T}$ in a top-down manner starting from root node A.
  - For each node x, set its degree as $d_x = z_x - 1$, where z_x is the threshold value of x.
  - For the polynomial q_A of the root node A, we randomly choose $q_A(0) \epsilon \mathbb{Z}_p$ as well as d_A additional points satisfy q_A. Furthermore, set $\sigma_A = q_A(0)$.
  - For the other nodes x, $q_x(0)$ is computed as $q_x(0) = q_{parent(x)}(index(x))$ ($parent(x)$ is parent node of x, $index(x)$ represents that the order number of x in all children). d_x other points of polynomial q_x are chosen randomly to define it completely. Set $\sigma_x = q_x(0)$.
  - For each leaf node y, compute: $C_y = h^{q_y(0)}, C_y' = H_1(att(y))^{q_y(0)}$.
  - Compute: $C_x^{1'} = e(g, g)^{\alpha\sigma_x}, C_x^2 = g^{\sigma_x}$.

- **OnlineEnc** $(Param, \mathcal{T}, \{C_X^1\}_{x \in \mathcal{T}}, \{M_x\}_{x \in X}) \rightarrow (\{C_X^1\}_{x \in \mathcal{T}}, \{C_X^3\}_{x \in \mathcal{T}})$
  - For each $x \in X$, choose a random: $R_x \in Z_p$. X is the set of x that will be used in the encryption.
  - Compute: $C_x^1 = R_x \cdot C_x^{1'}, k_x = H_2(C_x^1 \parallel C_x^2 \parallel R_x), C_x^3 = E_{k_x}(M_x)$.

Finally, PO sends the $CT = \{\mathcal{T}, \{C_x^1, C_x^2, C_x^3\}_{\forall x \in X}, \{C_y, C_y',\}_{\forall y \in Y}\}$ to the CSP for sharing.

Re-encryption. To achieve attribute-level user revocation, the CSP needs to re-encrypt the ciphertext so that only legitimate users can decrypt the ciphertext. It needs to generate and update the secret key for the attribute group. Using the UL, it selects the root nodes with the minimum cover sets in the KEK tree. Such a selection should cover all leaf nodes associated with the users in G_i. The group key of the attribute is denoted as $KEK(G_i)$. For example, if $G_i = (u_1, u_3, u_7, u_8)$ in Fig. 4, then $KEK(G_i) = \{KEK_8, KEK_{10}, KEK_7\}$ because v_8, v_{10} and v_7 are the root nodes with the minimum cover sets for all members in G_i. Therefore, it covers all users in G_i. Users not in G_i have no knowledge of KEK in $KEK(G_i)$. Once CT is received from the PO, CSP first randomly chooses $\delta_i \in Z_P$, and computes $C_y'' = C_y'^{1/\delta_i}$. Then, a header message $Hdr = (\forall y \in Y : \{E_K(\delta_i)\}_{K \in KEK(G_y)})$ is generated, where $E_K(M)$ is an encryption of M using the key K. Finally, the CSP stores the ciphertext $CT = \{Hdr, \mathcal{T}, \{C_x^1, C_x^2, C_x^3\}_{\forall x \in X}, \{C_y, C_y''\}_{\forall y \in Y}\}$.

PHR Access. After the PU requests to obtain the ciphertext, it needs to perform two steps to recover the PHRs, decryption for the attribute group key and decryption for the message. When PU receives the ciphertext (Hdr, CT'), he first obtains the attribute group keys for all attributes which will be used from Hdr. If he is the valid user in G_i, he has a valid KEK which can be used in the symmetric decryption. For example, if $G_i = (u_1, u_3, u_7, u_8)$ in Fig. 4, then $KEK(G_i) = \{KEK_8, KEK_{10}, KEK_7\}$. The PK of u_8 is $\{KEK_1, KEK_3, KEK_7, KEK_{15}\}$. The only common key is KEK_7, then u_8 can decrypt Hdr for the group key. The PK of u_2 is $\{KEK_1, KEK_2, KEK_4, KEK_9\}$ which can not be used for a valid key. PU can compute its decryption key with the attribute group keys as follows: $\{D = g^\alpha h^r, \{D_i = g^r \cdot H_1(i)^{r_i}, D_i' = h^{r_i \delta_i}\}, \forall i \in \gamma\}$

Then, PU Performs the decrypt algorithm which is defined in the same way as [28] with the decryption key, and the message can be recovered as follows:

- For each leaf node y, let $i = attt(y)$, compute:

$$DecNode(y) = \frac{e(D_i, C_y)}{e(D_i', C_y')} = \frac{e(g^r \cdot H_1(i)^{r_i}, h^{q_y(0)})}{e(h^{r_i \delta_i}, H_1(i)^{(1/\delta_i) \cdot q_y(0)})} = e(g, g)^{r \beta q_y(0)} \quad (2)$$

- For each non-leaf node x, compute the algorithm DecNode(x) recursively:

$$
\begin{aligned}
DecNode(x) &= \prod_{x' \in S_x} DecNode(x')^{\Delta_{j, S_x'}(0)} \\
&= \prod_{x' \in S_x} (e(g, g)^{r \beta q_{x'}(0)})^{\Delta_{j, S_x'}(0)} \\
&= \prod_{x' \in S_x} (e(g, g)^{r \beta q_{parent(x')} t(indext(x'))})^{\Delta_{j, S_x'}(0)} \quad (3) \\
&= \prod_{x' \in S_x} e(g, g)^{r \beta q_x(j) \cdot \Delta_{j, S_x'}(0)} \\
&= e(g, g)^{r \beta q_x(0)}
\end{aligned}
$$

- Compute:

$$\frac{C_x^1}{e(C_x^2, D)/DecNode(x)} = \frac{R_x \cdot e(g, g)^{\alpha \sigma_x}}{e(g, g)^{\alpha \sigma_x}} = R_x \quad (4)$$

- Compute:

$$k_x = H_2(C_x^1 \parallel C_x^2 \parallel R_x), M_x = D_{k_x}(C_x^3) \tag{5}$$

Finally, the download of PU is recorded on the blockchain as a transaction record for review by other parties.

Attribute Revocation. If the list of revoked RL is changed, somebody leaving or joining, and TA needs to upload the latest RL to the blockchain. The CSP can perform as follows for CT updating and key updating.

- It randomly selects $s' \in Z_p$, and a different group key δ_i'.
- The ciphertext CT is re-encrypted. For $T, \forall x \in X$, it computes $C_x^1 = R_x \cdot e(g,g)^{\alpha\sigma_x + s'}, C_x^2 = g^{\sigma_x + s'}, C_x^3 = E_{k_x}(M_X)$, $C_i = h^{q_i(0)+s'}$, $C_i' = (H_1(att(y))^{q_i(0)+s'})^{\frac{1}{\delta_i}}$. For $T, \forall y \in Y/\{i\}$, it computes $C_y = h^{q_y(0)+s'}, C_y' = (H_1(att(y))^{q_y(0)+s'})^{1/\delta_i}$.
- It selects new minimum cover sets for G_i, and sets the group key is $KEK(G_i)$. And it computes $Hdr = (\{E_K(\delta_i')\}_{K \in KEK(G_i)}, \forall y \in Y/\{i\} : \{E_K(\delta_i)\}_{K \in KEK(G_y)})$.

6   Security Analyses

In this section, we prove that the proposed scheme satisfies the desired security requirements, based on the CBDH assumption.

Theorem 1 *Our proposed scheme is IND-CCA secure in the random oracle model based on the CBDH assumption and assuming the symmetric cipher (E,D) is IND-CCA secure. (Note that the protocol itself is only CPA secure, but using Fujisaki-Okamoto conversion, it can be made CCA secure.)*

Proof. Recall that $\mathcal{A}$ is a PPT adversary, who makes q_k key queries and q_d decryption Queries in time t, can win the IND-CCA game in our scheme with an advantage ϵ. B is a simulator that is used to solve the CBDH problem, and it controls H_1 and H_2. With the given instance (g, g^a, g^b, g^c), they interact with each other as follows:

- **Setup.** B performs the simulated system initialization algorithm to generate a set of parameters. According to the instance, it first chooses a random $\alpha' \in Z_p$, and sets $\alpha = \alpha' + ab$, $\beta = b$, computing $e(g,g)^\alpha = e(g,g)^{\alpha' + ab} = e(g,g)^{\alpha'} e(g,g)^{ab}$. The public parameter $Param = \{G_0, G_T, e, H_1, H_2, g, h = g^b, e(g,g)^{\alpha'} e(g,g)^{ab}, (E,D)\}$. It constructs a KEK tree with k leaf nodes which means that at most k different users can be asked for their keys.
  H_1 queries: For each H_1 query of attribute i, if there is a i to make $H_1(i) = s_i$ in the H_1 list, it returns the corresponding value; otherwise, it chooses a random $s \in G_0$ as the return value and adds the $(i, s_i = s)$ to the list.
  H_2 queries: For each H_2 query, if there is a (m, v_m) in the H_2 list, it returns the corresponding value v_m, otherwise it choose a random value v_m from $\{0,1\}^l$ as the result and add the (m, v_m) to the list.

- **Phase 1.** In this phase, the adversary $\mathcal{A}$ performs two kinds of queries as follows:

  Key Queries: $\mathcal{A}$ makes key queries for user with attribute set I in an adaptive way. B randomly chooses $r' \in Z_p$ and sets $r = r' - a$, computing:
  $D = g^\alpha h^r = g^{\alpha' + ab}g^{b(r'-a)} = g^{\alpha' + br'}$, $D_i = \frac{g^{r'}}{g^a} \cdot H_1(i)^{r_i}$. It chooses path keys from the KEK tree as PK for the user. Note that the PK was not distributed to others and the different user has different PK. Add the user to the user list. The decryption keys $SK = \{D = g^{\alpha' + br'}, \forall i \in I : D_i = \frac{g^{r'}}{g^a} \cdot H_1(i)^{r_i}, D_i' = h^{r_i}\}$, $PK = \{KEK(i)\}_{i \in path}$.

  Decryption Queries: $\mathcal{A}$ requests the decryption query of the ciphertext $CT = \{T, C^1, C^2, C^3, C_y, C_y', Hdr\}$. B first decrypts the Hdr with the KEK_i in the tree, if there is no δ_i that can be found, it outputs $\perp$. Then, B searches hash list H_1, H_2 for the value (i, s_i) and (m, v_m). It extracts R from $v_i = C^1 \parallel C^2 \parallel R$ and checks if the pairs (i, s_i), and (m, v_m) satisfy the following equations: $K = V_i = H_2(C^1 \parallel C^2 \parallel R)$, $M = D_K(C^3)$, $C^1 = R \cdot e(g,g)^{\alpha' \sigma}e(g,g)^{ab\sigma}$, $C^2 = g^\sigma$, $C_y' = s_i^{q_y(0)\frac{1}{\delta_i}}$. If there exist hash queries satisfying the equations:

  $$DecNode(y) = \frac{e(D_i, C_y)}{e(D_i', C_y')} = \frac{e(\frac{g^{r'}}{g^a} \cdot H_1(i)^{r_i}, h^{q_y(0)})}{e(h^{r_i\delta_i}, s_i^{q_y(0)\frac{1}{\delta_i}})} = (\frac{e(g,g)^{r'b}}{e(g,g)^{ab}})^{q_y(0)} \quad (6)$$

  $$DecNode(x) = (\frac{e(g,g)^{r'b}}{e(g,g)^{ab}})^\sigma \quad (7)$$

  B returns the result to $\mathcal{A}$, otherwise it outputs $\perp$.
- **Challenge.** $\mathcal{A}$ chooses a challenge access structure T^* for the the user u_i. It selects two messages M_0, M_1 with the same length. For $\forall y \in Y$, it chooses a random number $\delta_i \in Z_p$, and computes the Hdr that $Hdr = (\forall y \in Y : \{E_K(\delta_i)\}_{K \in KEK(G_y)})$, where $EKE(G_j)$ is not available for u_i. Then, B randomly chooses $R_1 \in G_T$, $R_3, R_4 \in G_0$, and R_2 from the symmetric cipher's ciphertext space. Then, $CT^* = \{T^*, R_1, g^c, R_2, R_3, R_4\}$ is computed. If $\mathcal{A}$ does not query $att(y)$, with the random number σ^*, it is the correct CT^* which can be seen as $R_1 = R \cdot e(g,g)^{\alpha'c}e(g,g)^{abc}$, $R_2 = E_K(M_\theta)$, $R_3 = h^{q_y(0)}$, $R_4 = H_1(att(y))^{q_y(0)\delta_i}$.
- **Phase 2.** In this phase, $\mathcal{A}$ can make queries in the same way as in phase 1 with two exceptions: First, it cannot query the access structure that matches T^* in the key query. And second, it cannot query the challenge ciphertext CT^* in the decryption query.
- **Guess.** $\mathcal{A}$ guesses the bit θ or outputs $\perp$.

Discussion: For all H_2 queries, the simulator chooses a random one as the challenge hash query which is defined as $H_2^* = C^1 \parallel C^2 \parallel R = R \cdot e(g,g)^{\alpha'c}e(g,g)^{abc} \parallel g^c \parallel R$. In this way, if the H_2^* is used for the challenge by $\mathcal{A}$, the simulator can use it to solve the CBDH problem. In the above game, the decryption simulation is correct except with a negligible probability:

– According to the description in *Phase 1*, B can answer the decryption query on $CT = \{T, C^1, C^2, C^3, C_y, C'_y, Hdr\}$.
– If there is a required CT which satisfy $C^2 = g^c$ (g^c is the challenge ciphertext). B must return $\perp$ to $\mathcal{A}$. Because B can determine the corresponding ciphertext, and the CT which is different from CT^* is an invalid ciphertext.
– If $C^2 \not\equiv g^c$, but $\mathcal{A}$ has queried (i, s_i) and (m, v_m) which the challenge M_θ can be satisfied as following:

$$K = V_i = H_2(C^1 \parallel C^2 \parallel R), M_\theta = D_K(C^3),$$

$$C^1 = R \cdot e(g,g)^{\alpha' \sigma} e(g,g)^{ab\sigma}, C^2 = g^\sigma, C'_y = s_i^{q_y(0)\frac{1}{\delta_i}}$$

$$DecNode(y) = \frac{e(D_i, C_y)}{e(D'_i, C'_y)} = \left(\frac{e(g,g)^{r'b}}{e(g,g)^{ab}}\right)^{q_y(0)}, DecNode(x) = \left(\frac{e(g,g)^{r'b}}{e(g,g)^{ab}}\right)^\sigma$$

In this situation, B returns the corresponding result to $\mathcal{A}$. However, it has a negligible probability to hold this kind of equation, because s_i and v_m are selected randomly. Furthermore, $(a, b, r, r_i, R, s_i, v_m, R_1, R_2, R_3, R_4)$ are chosen randomly in the simulation game so that the simulation is indistinguishable in the view of adversary $\mathcal{A}$.

7   Efficiency Analyses

In this part, we analyze the performance of our RH-ABE with related works in this field, TRAC in [17], data sharing scheme in [24], TR-AP-CPABE in [10].

Features. As shown in Table 1, We compare their features in seven dimensions: Access Model(AC), Hierarchical Access Control(HAC), Offline/Online Encryption(Off/On En), Traceable(Tra), User Revocation(UR), Attribute-level Revocation(Attr UR), Security Level(SL). Li et al. [17] proposed a traceable and revocable access control scheme for mHealth which is an And gate access structure scheme with IND-CPA security. Tan et al. [24] proposed a data-sharing scheme for COVID-19 Medical Records. It is a blockchain-empowered approach that can achieve malicious user traceability and revocation. It is also IND-CPA secure. Han et al. [10] proposed a traceable and revocable ABE scheme with a hidden policy. Both of them are user revocation schemes with IND-CPA, and they can not achieve attribute-level revocation. In our scheme, we proposed a hierarchical access control scheme with offline/online encryption and attribute-level user revocation which can achieve IND-CCA security. In terms of features, our scheme is more suitable for mobile health systems.

7.1   Computation Costs

In this part, we ignore the computational costs of operations in symmetric cipher as well as hash functions, as they are much lighter compared with the costs in

Table 1. Comparison of Features.

Work	AM	HAC	Off/On En	Tra	UR	Attr UR	SL
[17]	And gate	No	No	Yes	Yes	No	$IND - CPA$
[24]	Tree	No	No	Yes	Yes	No	$IND - CPA$
[10]	Tree	No	No	Yes	Yes	No	$IND - CPA$
Ours	Tree	Yes	Yes	No	Yes	Yes	$IND - CPA$

attribute-based encryption. Table 2 shows the number of exponential and paring operations in the four schemes. E represents an exponential operation and P represents a bilinear pair operation. $|U|$ is the number of users in the system. $|I|$ is the number of attributes in the system. $|R|$ is the max number of revocation user and $|S|$ represents a user's attribute number. r is the number of attributes in the access policy.

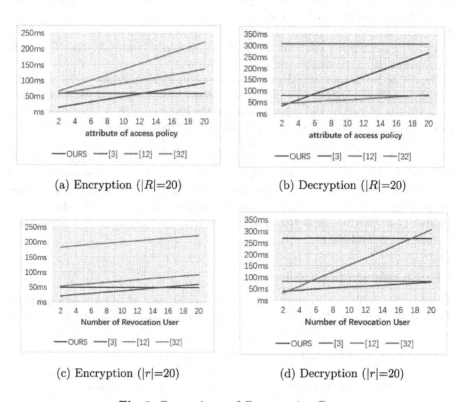

(a) Encryption ($|R|$=20)

(b) Decryption ($|R|$=20)

(c) Encryption ($|r|$=20)

(d) Decryption ($|r|$=20)

Fig. 5. Comparisons of Computation Cost

In the *Setup* phase, the computation costs in [17] and [10] are linear to the number of users, but they are constant in [24] and our scheme. The computation

Table 2. Comparison of Computation cost.

	Setup	KeyGen	Encrypt	Decrypt
[17]	$(4\|U\|+2)E+2P$	$(3+2\|I\|)E$	$(7+\|R\|)E$	$(4+\|R\|)E+4P$
[24]	$2E+P$	$(3+\|S\|)E$	$(2+r+\|R\|)E$	$\|R\|E+2\|R\|P$
[10]	$2\|U\|E+P$	$(4+2\|S\|)E$	$(2+4r+\|R\|)E$	$(3+r)E+5P$
Ours	$2E+P$	$(2+2\|S\|)E$	$(2X+2r)E$	$(1+2r)P$

cost of *KeyGen* in [17] is linear to the number of system attributes, but it relates to the number of user's attributes in [10,24] and our scheme. Considering that the user's attribute set is generally smaller than the system attribute set. The computation cost of encryption in [17] is related to the number of revocation users. The larger the user revocation list, is more computationally expensive. In our scheme, The encryption costs are related to the access structure. The larger the access policy, the higher computational cost in our scheme. In general, the size of the access policy does not change regularly, but the number of revoked users may increase. In [10,24], the encryption costs are related to the size of $|R|$ and r. The decryption costs in [17] and [24] are only related to the $|R|$, while the work in [10] and our scheme are related to r.

In Fig. 5, we graphically depict how the computational cost varies with the size of $|R|$ and r. The time of a pairing is 6.59 ms, and the time of exponentiation is 2.18 ms. In Fig. 5 (a) and (b), we assumed the number of revocation users is 20, and the computation cost of encryption and decryption varies with the number of attributes of the access policy. In Fig. 5(c) and (d), we assumed the size of r is 20, and the computation cost of encryption and decryption varies with the number of revocation users. In both cases, the computation encryption cost of our scheme is relatively lower. The decryption efficiency is not very good, since the pairing operation is related to the size of the access structure. But, as the number of revoked users in the system increases, the cost of other schemes will become larger and larger.

7.2 Communication Costs

As shown in Table 3, we compare the communication cost of each scheme from three aspects, size of the public parameter, secret key, and ciphertext.

Table 3. Comparison of Communication Cost.

	Size of Public Parameter	Size of Secret Key	Size of Ciphertext
[17]	$(3\|I\|+2\|U\|+2)G_0+G_T$	$(3+2\|I\|)G_0$	$(5+\|R\|)G_0+G_T$
[24]	$2G_0+G_T$	$(3+\|S\|)G_0$	$(1+r+\|R\|)G_0+G_T$
[10]	$(3+2\|U\|)G_0+G_T$	$(3+\|S\|)G_0$	$(2+3r+\|R\|)G_0+G_T$
Ours	$2G_0+G_T$	$(1+2\|S\|)E$	$(1+2r)G_0+G_T$

In the scheme [17], the size of the public parameter is related to the number of system users and system attributes. And public parameter in [10] is related to the number of system users. Both of them are more than $2G_0 + G_T$. The size of the secret key in [17] is related to the number of system attributes while others are related to the user's attributes. And communication cost of ciphertext is related to the size of revoked user list except for our scheme. Totally, the communication cost in our scheme is the least.

8   Conclusion

In this paper, a blockchain-based hierarchical access control with efficient revocation is proposed for the mHealth system. Our scheme is based on the file-related hierarchical access control scheme, we use the KEK technique for attribute-lever user revocation. PO can encrypt a series of data with hierarchical access control relationships at the same time to improve efficiency which is suitable for a resource-constrained environment. We also use the offline/online encryption model, since the access policy can be chosen in advance. Fine-grained user revocation is desirable because users' roles may change regularly. It is also proved that our scheme is IND-CCA secure. The comparison with some related works demonstrates that RH-ABE is generally efficient for the mHealth system.

Note that the pairing operation of decryption grows linearly with the scale of access policy, in our scheme. There are many schemes that outsource decryption computing to cloud providers. But some of them do not use hierarchical control, and others do not provide attribute-level user revocation. It makes sense to build solutions that are efficient in all aspects.

References

1. Bao, Y., Qiu, W., Tang, P., Cheng, X.: Efficient, revocable, and privacy-preserving fine-grained data sharing with keyword search for the cloud-assisted medical iot system. IEEE J. Biomed. Health Inform. **26**(5), 2041–2051 (2021)
2. Bethencourt, J., Sahai, A., Waters, B.: Ciphertext-policy attribute-based encryption. In: 2007 IEEE Symposium on Security and Privacy (SP'07), pp. 321–334. IEEE (2007)
3. Boneh, D., Franklin, M.: Identity-based encryption from the weil pairing. In: Kilian, J. (ed.) CRYPTO 2001. LNCS, vol. 2139, pp. 213–229. Springer, Heidelberg (2001). https://doi.org/10.1007/3-540-44647-8_13
4. Edemacu, K., Jang, B., Kim, J.W.: Collaborative ehealth privacy and security: An access control with attribute revocation based on OBDD access structure. IEEE J. Biomed. Health Inform. **24**(10), 2960–2972 (2020)
5. Fu, J., Wang, N.: A practical attribute-based document collection hierarchical encryption scheme in cloud computing. IEEE Access **7**, 36218–36232 (2019)
6. Goyal, V., Pandey, O., Sahai, A., Waters, B.: Attribute-based encryption for fine-grained access control of encrypted data. In: Proceedings of the 13th ACM Conference on Computer and Communications Security, pp. 89–98 (2006)
7. Green, M., Hohenberger, S., Waters, B.: Outsourcing the decryption of {ABE} ciphertexts. In: 20th USENIX Security Symposium (USENIX Security 11) (2011)

8. Guo, R., Yang, G., Shi, H., Zhang, Y., Zheng, D.: O 3-R-CP-ABE: an efficient and revocable attribute-based encryption scheme in the cloud-assisted iomt system. IEEE Internet Things J. **8**(11), 8949–8963 (2021)
9. Hajar, M.S., Al-Kadri, M.O., Kalutarage, H.K.: A survey on wireless body area networks: architecture, security challenges and research opportunities. Comput. Secur. **104**, 102211 (2021)
10. Han, D., Pan, N., Li, K.C.: A traceable and revocable ciphertext-policy attribute-based encryption scheme based on privacy protection. IEEE Transactions on Dependable and Secure Computing (2020)
11. Hur, J., Noh, D.K.: Attribute-based access control with efficient revocation in data outsourcing systems. IEEE Trans. Parallel Distrib. Syst. **22**(7), 1214–1221 (2010)
12. Lee, K., Choi, S.G., Lee, D.H., Park, J.H., Yung, M.: Self-updatable encryption: time constrained access control with hidden attributes and better efficiency. In: Sako, K., Sarkar, P. (eds.) Advances in Cryptology - ASIACRYPT 2013, pp. 235–254. Springer Berlin Heidelberg, Berlin, Heidelberg (2013). https://doi.org/10.1007/978-3-642-42033-7_13
13. Li, H., Yu, K., Liu, B., Feng, C., Qin, Z., Srivastava, G.: An efficient ciphertext-policy weighted attribute-based encryption for the internet of health things. IEEE J. Biomed. Health Inform. **26**(5), 1949–1960 (2021)
14. Li, J., Yu, Q., Zhang, Y.: Hierarchical attribute based encryption with continuous leakage-resilience. Inf. Sci. **484**, 113–134 (2019)
15. Li, M., Lou, W., Ren, K.: Data security and privacy in wireless body area networks. IEEE Wirel. Commun. **17**(1), 51–58 (2010)
16. Li, M., Yu, S., Zheng, Y., Ren, K., Lou, W.: Scalable and secure sharing of personal health records in cloud computing using attribute-based encryption. IEEE Trans. Parallel Distrib. Syst. **24**(1), 131–143 (2012)
17. Li, Q., Xia, B., Huang, H., Zhang, Y., Zhang, T.: Trac: traceable and revocable access control scheme for mhealth in 5G-enabled iiot. IEEE Trans. Industr. Inf. **18**(5), 3437–3448 (2021)
18. Liang, X., Lu, R., Lin, X., Shen, X.S.: Ciphertext policy attribute based encryption with efficient revocation. TechnicalReport, University of Waterloo **2**(8) (2010)
19. Nishide, T., Yoneyama, K., Ohta, K.: Attribute-based encryption with partially hidden encryptor-specified access structures. In: Bellovin, S.M., Gennaro, R., Keromytis, A., Yung, M. (eds.) Applied Cryptography and Network Security: 6th International Conference, ACNS 2008, New York, NY, USA, June 3-6, 2008. Proceedings, pp. 111–129. Springer Berlin Heidelberg, Berlin, Heidelberg (2008). https://doi.org/10.1007/978-3-540-68914-0_7
20. Riad, K., Huang, T., Ke, L.: A dynamic and hierarchical access control for Iot in multi-authority cloud storage. J. Netw. Comput. Appl. **160**, 102633 (2020)
21. Sahai, A., Waters, B.: Fuzzy identity-based encryption. In: Cramer, R. (ed.) EUROCRYPT 2005. LNCS, vol. 3494, pp. 457–473. Springer, Heidelberg (2005). https://doi.org/10.1007/11426639_27
22. Shen, J., Gui, Z., Chen, X., Zhang, J., Xiang, Y.: Lightweight and certificateless multi-receiver secure data transmission protocol for wireless body area networks. IEEE Transactions on Dependable and Secure Computing (2020)
23. Sun, J., Xiong, H., Liu, X., Zhang, Y., Nie, X., Deng, R.H.: Lightweight and privacy-aware fine-grained access control for Iot-oriented smart health. IEEE Internet Things J. **7**(7), 6566–6575 (2020)
24. Tan, L., Yu, K., Shi, N., Yang, C., Wei, W., Lu, H.: Towards secure and privacy-preserving data sharing for Covid-19 medical records: a blockchain-empowered approach. IEEE Trans. Netw. Sci. Eng. **9**(1), 271–281 (2021)

25. Tang, W., Zhang, K., Ren, J., Zhang, Y., Shen, X.: Lightweight and privacy-preserving fog-assisted information sharing scheme for health big data. In: GLOBE-COM 2017-2017 IEEE Global Communications Conference, pp. 1–6. IEEE (2017)
26. Wang, S., et al.: A fast CP-ABE system for cyber-physical security and privacy in mobile healthcare network. IEEE Trans. Ind. Appl. **56**(4), 4467–4477 (2020)
27. Wang, S., Zhou, J., Liu, J.K., Yu, J., Chen, J., Xie, W.: An efficient file hierarchy attribute-based encryption scheme in cloud computing. IEEE Trans. Inf. Forensics Secur. **11**(6), 1265–1277 (2016)
28. Xiao, M., Li, H., Huang, Q., Yu, S., Susilo, W.: Attribute-based hierarchical access control with extendable policy. IEEE Transactions on Information Forensics and Security (2022)
29. Xu, S., Li, Y., Deng, R., Zhang, Y., Luo, X., Liu, X.: Lightweight and expressive fine-grained access control for healthcare internet-of-things. IEEE Transactions on Cloud Computing (2019)
30. Yang, Y., Liu, X., Deng, R.H., Li, Y.: Lightweight sharable and traceable secure mobile health system. IEEE Trans. Dependable Secure Comput. **17**(1), 78–91 (2017)
31. Zhang, L., Zhao, C., Wu, Q., Mu, Y., Rezaeibagha, F.: A traceable and revocable multi-authority access control scheme with privacy preserving for mhealth. J. Syst. Architect. **130**, 102654 (2022)
32. Zhong, H., Zhou, Y., Zhang, Q., Xu, Y., Cui, J.: An efficient and outsourcing-supported attribute-based access control scheme for edge-enabled smart healthcare. Futur. Gener. Comput. Syst. **115**, 486–496 (2021)

Author Index

© ICST Institute for Computer Sciences, Social Informatics and Telecommunications Engineering 2024
Published by Springer Nature Switzerland AG 2024. All Rights Reserved
J. Chen and Z. Xia (Eds.), BlockTEA 2023, LNICST 577, p. 155, 2024.
https://doi.org/10.1007/978-3-031-60037-1

Printed in the United States
by Baker & Taylor Publisher Services